Praise for

THE WISDOM OF PSYCHOPATHS

"It's hard not to like Dutton's book . . . Dutton, like [Norman] Mailer, is waging war against the *bien-pensant.* And I'm with him. Life would be more fun if more people cultivated their inner psychopath." —Ann Marlowe, *Tablet*

"Dutton deftly navigates through some disturbing subject matter, but his message is ultimately upbeat: Scientists may be able to learn a lot from the darker side of human nature."

—Allison Bohac, *Science News*

"A convincing study . . . The admirable quality of this book is Dutton's refusal to accept easy answers in one of the more sensational fields of popular psychology."

—Tim Adams, *The Observer* (U.K.)

"Dutton spins a solid yarn, turning what could easily have been a dry survey of psych research into entertainment."

—Scott Olster, *Fortune*

"An enjoyable, breezy treatment of a provocative subject."

—*Kirkus Reviews*

"*The Wisdom of Psychopaths* is a surprising, absorbing, and perceptive book. Kevin Dutton has amassed a great deal of knowledge about these charming, cold, fearless, emotionally indifferent people, who are so attractive in some ways and so appalling in others, and set it out in briskly readable prose studded with gripping anecdotes. I found it altogether fascinating."

—Philip Pullman, author of the bestselling *His Dark Materials* trilogy

"Dutton tackles an elusive, important, and much neglected aspect of the mind: our personality. He presents some highly original insights and does so in a provocative and humorous style—offering practical tips along the way for both 'normals' and 'sociopaths.'"

—V. S. Ramachandran, Ph.D., author of *The Tell-Tale Brain: A Neuroscientist's Quest for What Makes Us Human*

"Dutton has written a masterful, readable, and entertaining treatise on psychopathy and its manifestations in everyday life. Some of his ideas will generate debate and controversy, but he clearly has provided a thought-provoking book for those seeking to understand the 'psychopathic' world in which we live."

—Robert D. Hare, Ph.D., author of *Without Conscience: The Disturbing World of the Psychopaths Among Us* and developer of the Hare Psychopathy Checklist

"The irrepressible Kevin Dutton has done it again! This time he has produced an irreverent romp through the bright side and the dark side of the mysterious psychopath, and does a great job of mixing the scientific with the personal, offering readers an insider's glimpse into the workings of fascinating persons—and fascinating personalities. Readers will come away both enlightened and entertained."

—Scott O. Lilienfeld, Ph.D., Professor of Psychology at Emory University, President of the Society for the Scientific Study of Psychopathy, and coauthor of *50 Great Myths of Popular Psychology*

"If you've been keeping your inner psychopath locked up in the maximum-security unit of your mind, Kevin Dutton explains why giving him some fresh air from time to time may actually do you—and, more important, the rest of us—a world of good. Just give him this book to read and make sure he's a literate, functional psychopath."

—Jesse Bering, Ph.D., author of *Why Is the Penis Shaped Like That?: And Other Reflections on Being Human* and *Perv: The Sexual Deviant in All of Us*

PAUL ROBERT WILLIAMS

KEVIN DUTTON

THE WISDOM OF PSYCHOPATHS

Dr. Kevin Dutton is a research psychologist at the Calleva Research Centre for Evolution and Human Sciences, Magdalen College, University of Oxford. He is a fellow of the Royal Society of Medicine and the Society for the Scientific Study of Psychopathy. Dutton is the author of *Split-Second Persuasion.* His writing and research have been featured in *Scientific American Mind*, *New Scientist*, *The Guardian*, *Salon*, *The Huffington Post*, *Psychology Today*, and *USA Today.* He lives in Oxford, England. Read more about Dutton at www.kevindutton.co.uk. Follow him at @profkevindutton.

ALSO BY KEVIN DUTTON

Split-Second Persuasion:
The Ancient Art and New Science of Changing Minds

THE WISDOM OF PSYCHOPATHS

THE WISDOM OF PSYCHOPATHS

WHAT SAINTS, SPIES, AND SERIAL KILLERS CAN TEACH US ABOUT SUCCESS

KEVIN DUTTON

SCIENTIFIC AMERICAN / FARRAR, STRAUS AND GIROUX
NEW YORK

Scientific American / Farrar, Straus and Giroux
18 West 18th Street, New York 10011

Printed in the United States of America
Published in 2012 by Scientific American / Farrar, Straus and Giroux
First paperback edition, 2013

An excerpt from chapter 1 originally appeared, in slightly different form, in *Scientific American*.

Grateful acknowledgment is made for permission to reprint excerpts from the following previously published material:
Eysenck's personality model: with kind permission from Springer Science+Business Media.
"The Lesson of the Moth," from *Archy and Mehitabel* by Don Marquis, copyright 1927 by Doubleday, a division of Random House, Inc. Used by permission of Doubleday, a division of Random House, Inc.

The Library of Congress has cataloged the hardcover edition as follows:
Dutton, Kevin.
The wisdom of psychopaths : what saints, spies, and serial killers can teach us about success / Kevin Dutton. — 1st ed.
p. cm.
Includes bibliographical references and index.
ISBN 978-0-374-29135-8 (alk. paper)
1. Success. 2. Psychopaths. 3. Neurosciences—Research. I. Title.
BF637.S8 D88 2012
155.2'32—dc23

2012021530

Paperback ISBN: 978-0-374-53398-4

Designed by Abby Kagan

www.fsgbooks.com
www.twitter.com/fsgbooks • www.facebook.com/fsgbooks
books.scientificamerican.com

7 9 10 8 6

In memory of John Richard Dutton

The mind is its own place, and in itself
Can make a heav'n of hell, a hell of heav'n.

—*PARADISE LOST* 1: 254-255

CONTENTS

PREFACE

My father was a psychopath. It seems a bit odd saying that now, looking back. But he was. No question. He was charming, fearless, ruthless (but never violent). And he had about as much going on in the conscience department as a Jeffrey Dahmer freezer. He didn't kill anyone. But he certainly made a few killings.

It's a good thing genes aren't everything, right?

My father also had an uncanny knack for getting exactly what he wanted, often with just a casual throwaway line or a single telling gesture. People used to say that he looked like the scheming Del Boy in *Only Fools and Horses*—which he did—not just acted like him, which he also did (he, too, was a market trader).

The BBC sitcom could have been a Dutton family video.

I once remember helping Dad sell a load of datebooks at Petticoat Lane Market, in London's East End. I was ten at the time, and it was a school day. The datebooks in question were a collector's item. They only had eleven months.

"You can't sell these," I protested. "There's no January!"

"I know," he said. "That's why I forgot your birthday."

"Unique opportunity to get your hands on an eleven-month diary, folks . . . sign up for a special two-for-one offer and get an extra month thrown in next year for free . . ."

We unloaded the whole damn lot.

I've always maintained that Dad was in possession of pretty much the ideal personality for modern living. I never once saw him panic. Never once saw him lose his cool. Never once saw him get hot under the collar about anything. And believe me, there were plenty of times when he might have.

"They say that humans developed fear as a survival mechanism to protect against predators," he once told me. "But you don't see too many saber-toothed tigers prowling round Elephant and Castle, now do you, boy?"

He was right. I certainly hadn't seen any. There were a few snakes, maybe. But everyone knew who *they* were.

For a long time, growing up, I used to think of Dad's bon mot as just another of his market-stall one-liners—here today, gone tomorrow. A bit like a lot of the crap he used to sell, funnily enough. But now, years later, I realize that there was actually a deep biological truth to what the crafty old guy was saying. In fact, he anticipated the position taken by modern evolutionary psychologists with uncanny, sublime precision. We humans, it appears, did indeed develop our fear response as a survival mechanism to protect against predators. Monkeys with lesions of the amygdala, for instance—the brain's emotional sorting office—do very stupid things, like trying to pick up cobras.

But millions of years on, in a world where wild animals aren't lurking around every street corner, this fear system can be oversensitive—like a nervous driver with a foot hovering constantly over the brake pedal—reacting to dangers that don't actually exist and pushing us into making illogical, irrational decisions.

"There was no such thing as stock in the Pleistocene era," George Loewenstein, professor of economics and psychology at Carnegie Mellon, points out. "But human beings are pathologically risk averse. A lot of the mechanisms that drive our emotions aren't really that well adapted to modern life."

I prefer my dad's version.

The observation that modern-day humans are pathologically risk averse does not, needless to say, mean that this has always been the case. In fact, it might even be argued that those of us today who are *clinically* risk averse—those of us, for instance, who suffer from chronic anxiety—simply have too much of a good thing. During the time of our ancestors, the existence of individuals who were hypervigilant to threat may well, evolutionary biologists suggest, have been decisive in the fight against predators—and from this point of view, anxiety would undoubtedly have served as a considerable adaptive advantage. The more sensitive you were to rustlings in the undergrowth, the more likely you'd have been to have kept yourself, your family, and your extended group members alive. Even today, anxious individuals are better than the rest of us at detecting the presence of threat: slip an angry face in among a display of happy or neutral faces on a computer screen and anxious people are far faster at picking it out than those who are non-anxious—not a bad ability to fall back on should you happen to find yourself alone at night and wandering around an unfamiliar neighborhood. Being anxious can sometimes be useful.

The notion that a mental disorder can occasionally come in handy—conferring extraordinary, outlandish advantages, as well as inordinate distress, on its sufferers—is hardly new, of course. As the philosopher Aristotle observed more than 2,400 years ago, "There was never a genius without a tincture of madness." In most people's minds, this link between "genius" and "madness" is probably most apparent, thanks to the box-office success of the films *Rain Man* and *A Beautiful Mind*, when it comes to autism and schizophrenia. In his book *The Man Who Mistook His Wife for a Hat*, the neurologist and psychiatrist Oliver Sacks reports a famous encounter with "the twins." Profoundly autistic, John and Michael, then twenty-six, were living in an institution. When a box of matches spilled onto the floor, both of them simultaneously called out "111." As Sacks gathered up the matches, he started counting . . .

On a similar note, the well-worn stereotype of the brilliant "tortured artist" is also not without foundation. The painter Vincent van Gogh, the dancer Vaslav Nijinsky, and the father of "game theory" (of which more later) John Nash were all psychotic. Coincidence? Not according to Szabolcs Kéri, a researcher at Semmelweis University in Budapest, who appears to have uncovered a genetic polymorphism associated with both schizophrenia and creativity. Kéri has found that people with two copies of a particular single-letter DNA variation in a gene called *neuregulin 1*, a variation that has been previously linked to psychosis—as well as poor memory and sensitivity to criticism—tend to score significantly higher on measures of creativity compared with individuals who have one or no copy of the variation. Those with one copy also tend to be more creative, on average, than those without.

Even depression has its advantages. Recent research suggests that despondency helps us think better—and contributes to increased attentiveness and enhanced problem-solving ability. In an ingenious experiment, Joe Forgas, professor of psychology at the University of New South Wales, placed a variety of trinkets, such as toy soldiers, plastic animals, and miniature cars, near the checkout counter of a small stationery store in Sydney. As shoppers made their way out, Forgas tested their memory, asking them to list as many of the items as possible. But there was a catch. On some days the weather was rainy, and Forgas piped Verdi's *Requiem* through the store; on other days it was sunny, and shoppers were treated to a blast of Gilbert and Sullivan.

The results couldn't have been clearer: shoppers in the "low mood" condition remembered nearly four times as many of the knickknacks. The rain made them sad, and their sadness made them pay more attention. Moral of the story? When the weather's nice, be sure to check your change.

When you go down the road of disorders conferring advantages, of clouds, silver linings, and psychological consolation prizes, it's difficult to conceive of a condition that *doesn't* pay off—at least in some form or another. Obsessive-compulsive? You're never going to leave the gas on. Paranoid? You'll never fall afoul of the small print. In fact, fear and sadness—anxiety and depression—constitute two of the five basic emotions* that are found universally across cultures, and that, as such, virtually all of us experience at some point in our lives. But

* The other three basic emotions are anger, happiness, and disgust. There is some dispute about the inclusion of a sixth, surprise, in the list.

there's one group of people who are the exception to the rule, who don't experience either—even under the most difficult and trying of circumstances. Psychopaths. A psychopath wouldn't worry even if he *had* left the gas on.* Any silver linings there?

Put this question to a psychopath and, more often than not, he'll look at you as if *you're* the one who's crazy. To a psychopath, you see, there are no such things as clouds. There are *only* silver linings. The fiendish observation that a year consists of twelve months, not eleven, might well have put one hell of a kibosh on selling those datebooks, you'd think. But not to my dad, it didn't. Quite the reverse, in fact. He saw it as a selling point.

He's certainly not alone. Nor, some might argue, is he too far off the mark. During the course of my research I've met a great many psychopaths from all walks of life—not just within my own family. Sure, behind closed doors I've encountered my fair share of Hannibal Lecters and Ted Bundys: remorseless, unconscionable A-listers who could dine at any psychopath table you care to mention without even picking up the phone—by just showing up. But I've also met psychopaths who, far from devouring society from within, serve, through nerveless poise and hard-as-nails decision making, to protect and enrich it instead: surgeons, soldiers, spies, entrepreneurs—dare I say, even lawyers. "Don't get too cocky. No matter how good you are. Don't let them see you coming," counseled Al Pacino as the head attorney of a top law firm in the film *The Devil's Advocate*. "That's the gaff, my friend—make yourself small. Be the hick. The cripple. The nerd. The leper. The freak. Look at me—I've been underestimated from day one." Pacino was playing the Devil. And—not surprisingly, perhaps—he hit the nail right on the head. If there's one thing that psychopaths have in common, it's the consummate ability to pass themselves off as normal everyday folk, while behind the facade—the brutal, brilliant disguise—beats the refrigerated heart of a ruthless, glacial predator.

* Most of the time, it *is* a "he." For the possible reasons why, see the notes section at the end of the book.

As one hugely successful young attorney told me on the balcony of his penthouse apartment overlooking the Thames: "Deep inside me there's a serial killer lurking somewhere. But I keep him amused with cocaine, Formula One, booty calls, and coruscating cross-examination."

Ever so slowly, I moved away from the edge.

This aerial encounter with the young lawyer (he later ran me back to my hotel downriver in his speedboat) goes some way toward illustrating a theory I have about psychopaths: that one of the reasons we're so fascinated by them is because we're fascinated by illusions, by things that appear, on the surface, to be normal, yet that on closer examination turn out to be anything but. *Amyciaea lineatipes* is a species of arachnid that mimics the physical appearance of the ants on which it preys. Only when it is too late are its victims finally disabused of the notion that they're good judges of character. Many people I've interviewed know exactly how that feels. And they, believe me, are the lucky ones.

Take a look at the picture below. How many soccer balls can you see? Six? Take another look. Still six? Turn to the end of the preface (page xix) and you'll find the answer at the bottom.

This is what psychopaths are like. Outwardly personable, they use their charm, charisma, and seamless psychological camouflage to distract us from their "true colors": the latent anomaly right in front of our eyes. Their intoxicating, hypnotic presence draws us inexorably in.

Yet psychopathy, as the Devil and his flamboyant London protégé just hinted, can also be good for us, at least in moderation. Like anxiety, depression, and quite a few other psychological disorders, it can at times be adaptive. Psychopaths, as we shall discover, have a variety of attributes—personal magnetism and a genius for disguise being just the starter pack—which, once you know how to harness them and keep them in check, often confer considerable advantages not just in the workplace, but in everyday life. Psychopathy is like sunlight. Overexposure can hasten one's demise in grotesque, carcinogenic fashion. But regulated exposure at controlled and optimal levels can have a significant positive impact on well-being and quality of life.

In the pages that follow we'll examine these attributes in detail, and learn how incorporating them into our own psychological skill set can dramatically transform our lives. Of course, it's in no way my intention to glamorize the actions of psychopaths—certainly not the actions of dysfunctional psychopaths, anyway. That would be like glamorizing a cognitive melanoma: the malignant machinations of cancer of the personality. But there's evidence to suggest that psychopathy, in small doses at least, is personality with a *tan*—and that it can have surprising benefits.

I've witnessed a few firsthand. As the years rolled by and he retired from the markets, the gods didn't look too favorably on Dad. (Though it wasn't as if he was picky: figurines of Buddhas, Muhammads, Sacred Hearts, Virgin Marys . . . they'd all done their time in the back of his three-wheeler van.) He got Parkinson's—and went, in a frighteningly short space of time, from someone who could pack up a suitcase in ten seconds flat (an ability that had come in handy surprisingly often) to someone who couldn't even stand without an aide on either arm ("In the old days, they used to be cops," he would say).

But his finest moment undoubtedly occurred posthumously. At

least, it was after he died that it came to my attention. One evening, not long after the funeral, I was going through his things when I came upon a volume of handwritten notes in a drawer. The notes had been penned by a succession of the various caregivers who'd looked after Dad over the previous few months (he'd managed, against the advice of pretty much everyone, to stick it out at home), and amounted, I suppose, to a kind of care "diary."

The first thing that struck me about the diary, I remember, was how neat and painstakingly detailed the entries were. Unmistakably female, the handwriting catwalked voluptuously across the page, modestly attired in blue or black Bic, with barely a serif or ligature out of place. But the more I read, the more it dawned on me just how little variety there had been in Dad's last few months: how monotonous, repetitive, and unremittingly bleak that final pitch, that final stand in the market stall of life, must've been. Not that he'd ever given me that impression when I'd dropped in to visit him, of course. The Parkinson's may well have been kicking the shit out of his arms and legs. But it was no match for his spirit.

Yet the reality of the situation was clear:

"Got Mr. Dutton out of bed at 7:30."

"Gave Mr. Dutton a shave."

"Made Mr. Dutton a cucumber sandwich."

"Brought Mr. Dutton a cup of tea."

And so on. And so forth. Ad infinitum.

Pretty soon I started to get bored, and, as one does, began randomly fanning through the pages. Then something caught my eye. In tremulous, spidery writing, scrawled in big block capitals across the middle of one of the pages, was the following: "MR. DUTTON DID CARTWHEELS DOWN THE HALL." Followed, a couple of pages later, by "MR. DUTTON PERFORMED A STRIP SHOW ON THE BALCONY."

Something told me he might be making it up. But hey, this was Dad we were talking about. Why mess with the habit of a lifetime?

Besides, the rules of the game had changed. Behind the cut-price bullshit lurked a higher, greater truth: the story of a man whose soul was under fire . . . whose circuits and synapses were hopelessly and mercilessly outgunned . . . but who, when the chips were down and the game was all but up, was going down fighting in a hail of irrepressible irreverence.

Cartwheels and strip shows beat shaves and cucumber sandwiches any day of the week.

Who cared if it was crap?

Okay. You're right, It *is* six. But now take a closer look at the man's hands. Notice anything unusual?

AUTHOR'S NOTE

The names and identifying details of certain people featured in this book have been changed. Such necessary demographic camouflage, however, does not in any way compromise the voice of these disguised individuals—and every step has been taken to report encounters and conversations as accurately and as authentically as possible. It should be noted that on account of the restrictions regarding recording equipment, this was especially the case in Broadmoor, where some degree of narrative license became inevitable in striking a balance between maintaining patient confidentiality and preserving the unique landscape of both the characters and the dialogue.

THE WISDOM OF PSYCHOPATHS

ONE

SCORPIO RISING

Great and Good are seldom the same man.

—WINSTON CHURCHILL

A scorpion and a frog are sitting on the bank of a river, and both need to get to the other side.

"Hello, Mr. Frog!" calls the scorpion through the reeds. "Would you be so kind as to give me a ride on your back across the water? I have important business to conduct on the other side. And I cannot swim in such a strong current."

The frog immediately becomes suspicious.

"Well, Mr. Scorpion," he replies, "I appreciate the fact that you have important business to conduct on the other side of the river. But just take a moment to consider your request. You are a scorpion. You have a large stinger at the end of your tail. As soon as I let you onto my back, it is entirely within your nature to sting me."

The scorpion, who has anticipated the frog's objections, counters thus:

"My dear Mr. Frog, your reservations are perfectly reasonable. But it is clearly not in my interest to sting you. I really do need to get to the other side of the river. And I give you my word that no harm will come to you."

The frog agrees, reluctantly, that the scorpion has a point. So he allows the fast-talking arthropod to scramble atop his back and hops, without further ado, into the water.

At first all is well. Everything goes exactly according to plan. But halfway across, the frog suddenly feels a sharp pain in his back—and sees, out of the corner of his eye, the scorpion withdraw his stinger from his hide. A deadening numbness begins to creep into his limbs.

"You fool!" croaks the frog. "You said you needed to get to the other side to conduct your business. Now we are both going to die!"

The scorpion shrugs and does a little jig on the drowning frog's back.

"Mr. Frog," he replies casually, "you said it yourself. I am a scorpion. It is in my nature to sting you."

With that, the scorpion and the frog both disappear beneath the murky, muddy waters of the swiftly flowing current.

And neither of them is seen again.

Bottom Line

During his trial in 1980, John Wayne Gacy declared with a sigh that all he was really guilty of was "running a cemetery without a license."

It was quite a cemetery. Between 1972 and 1978, Gacy had raped and murdered at least thirty-three young men and boys (with an average age of about eighteen) before stuffing them into a crawl space beneath his house. One of his victims, Robert Donnelly, survived Gacy's attentions, but was tortured so mercilessly by his captor that, at several points during his ordeal, he begged him to "get it over with" and kill him.

Gacy was bemused. "I'm getting around to it," he replied.

I have cradled John Wayne Gacy's brain in my hands. Following his execution in 1994 by lethal injection, Dr. Helen Morrison—a witness for the defense at his trial and one of the world's leading experts on serial killers—had assisted in his autopsy in a Chicago hospital, and then driven back home with his brain jiggling around in a glass jar on the passenger seat of her Buick. She'd wanted to find out whether there was anything about it—lesions, tumors, disease—that made it different from the brains of normal people.

Tests revealed nothing unusual.

Several years later, over coffee in her office in Chicago, I got to chatting with Dr. Morrison about the significance of her findings, the significance of finding . . . nothing.

"Does this mean," I asked her, "that we're basically all psychopaths deep down? That each of us harbors the propensity to rape, kill, and torture? If there's no difference between my brain and the brain of John Wayne Gacy, then where, precisely, does the difference lie?"

Morrison hesitated for a moment before highlighting one of the most fundamental truths in neuroscience.

"A dead brain is very different from a living one," she said. "Outwardly, one brain may look very similar to another, but function completely differently. It's what happens when the lights are on, not off, that tips the balance. Gacy was such an extreme case that I wondered whether there might be something else contributing to his actions—some injury or damage to his brain, or some anatomical anomaly. But there wasn't. It was normal. Which just goes to show how complex and impenetrable the brain can sometimes be, how reluctant it is to give up its secrets. How differences in upbringing, say, or other random experiences can cause subtle changes in internal wiring and chemistry which then later account for tectonic shifts in behavior."

Morrison's talk that day of lights and tectonic shifts reminded me of a rumor I once heard about Robert Hare, professor of psychology at the University of British Columbia and one of the world's leading authorities on psychopaths. Back in the 1990s, Hare submitted a research paper to an academic journal that included the EEG responses of both psychopaths and non-psychopaths as they performed what's known as a lexical decision task. Hare and his team of coauthors showed volunteers a series of letter strings, and then got them to decide as quickly as possible whether or not those strings comprised a word.

What they found was astonishing. Whereas normal participants identified emotionally charged words like "c-a-n-c-e-r" or "r-a-p-e" more quickly than neutral words like "t-r-e-e" or "p-l-a-t-e," this wasn't the case with psychopaths. To the psychopaths, emotion was irrelevant. The journal rejected the paper. Not it turned out, for its conclusions, but for something even more extraordinary. Some of the EEG patterns, reviewers alleged, were so abnormal they couldn't possibly have come from real people. But of course they had.

Intrigued by my talk with Morrison in Chicago about the mysteries and enigmas of the psychopathic mind—indeed, about neural recalcitrance in general—I visited Hare in Vancouver. Was the rumor true? I asked him. Had the paper really been rejected? If so, what was going on?

"There are four different kinds of brain waves," he told me, "ranging from beta waves during periods of high alertness, through alpha and theta waves, to delta waves, which accompany deep sleep. These waves reflect the fluctuating levels of electrical activity in the brain at various times. In normal members of the population, theta waves are associated with drowsy, meditative, or sleeping states. Yet in psychopaths they occur during normal waking states—even sometimes during states of increased arousal . . .

"Language, for psychopaths, is only word deep. There's no emotional contouring behind it. A psychopath may say something like 'I love you,' but in reality, it means about as much to him as if he said 'I'll have a cup of coffee.' . . . This is one of the reasons why psychopaths remain so cool, calm, and collected under conditions of extreme danger, and why they are so reward-driven and take risks. Their brains, quite literally, are less 'switched on' than the rest of ours."

I thought back to Gacy and what I'd learned from Dr. Morrison.

"Kiss my ass," he'd said as he entered the death chamber.

Normal on the outside (Gacy was a pillar of his local community, and on one occasion was even photographed with First Lady Rosalynn Carter), he camouflaged his inner scorpion with an endearing cloak of charm. But it was entirely in his nature to sting you—as much as it was to convince you that he wouldn't.

Talking the Walk

Fabrizio Rossi is thirty-five years old, and used to be a window cleaner. But his predilection for murder eventually got the better of him. And now, would you believe, he "does" it for a living.

As we stand next to each other on a balmy spring morning, poking uneasily around John Wayne Gacy's bedroom, I ask him what the deal is. What is it about psychopaths that we find so irresistible? Why do they fascinate us so much?

It's definitely not the first time he's been asked.

"I think the main thing about psychopaths," says Rossi, "is the fact that on the one hand they're so normal, so much like the rest of us—but on the other, so different. I mean, Gacy used to dress up as a clown and perform at children's parties . . . That's the thing about psychopaths. On the outside they seem so ordinary. Yet scratch beneath the surface, peek inside the crawl space, as it were, and you never know what you might find."

We are not, of course, in Gacy's actual bedroom, But rather, in a mocked-up version of it that comprises an exhibit in what must surely be a candidate for the grisliest museum in the world: the Museum of Serial Killers in Florence. The museum is located on Via Cavour, a ritzy side street within screaming distance of the Duomo.

And Fabrizio Rossi curates it.

The museum is doing well. And why wouldn't it? They're all there, if you're into that kind of thing. Everyone from Jack the Ripper to Jeffrey Dahmer, from Charles Manson to Ted Bundy.

Bundy's an interesting case, I tell Rossi. An eerie portent of the psychopath's hidden powers. A tantalizing pointer to the possibility that, if you look hard enough, there might be more in the crawl space than just *dark* secrets.

He's surprised, to say the least.

"But Bundy is one of the most notorious serial killers in history," he says. "He's one of the museum's biggest attractions. Can there really be anything else except dark secrets?"

There can. In 2009, twenty years after his execution at Florida State Prison (at the precise time that Bundy was being led to the electric chair, local radio stations urged listeners to turn off household appliances to maximize the power supply), psychologist Angela Book and her colleagues at Brock University in Canada decided to take the icy serial killer at his word. During an interview, Bundy, who staved in the skulls of thirty-five women during a four-year period in the mid-1970s, had claimed, with that boyish, all-American smile of his, that he could tell a "good" victim simply from the way she walked.

"I'm the coldest son of a bitch you'll ever meet," Bundy enunciated. And no one can fault him there. But, Book wondered, might he also have been one of the shrewdest?

To find out, she set up a simple experiment. First, she handed out the Self-Report Psychopathy Scale—a questionnaire specifically designed to assess psychopathic traits within the general population, as opposed to within a prison or hospital setting—to forty-seven male undergraduate students. Then, based on the results, she divided them up into high and low scorers. Next, she videotaped the gait of twelve new participants as they walked down a corridor from one room to another, where they completed a standard demographics questionnaire. The questionnaire included two items: (1) Have you ever been victimized in the past (yes or no)? (2) If yes, how many times has such victimization occurred?

Finally, Book presented the twelve videotaped segments to the original forty-seven participants, and issued them a challenge: rate, on a scale of one to ten, how vulnerable to being mugged each of the targets was. The rationale was simple. If Bundy's assertion held water and he really had been able to sniff out weakness from the way his victims walked, then, Book surmised, those who scored high on the Self-Report Psychopathy Scale should be better at judging vulnerability than the low scorers.

That, it turned out, was exactly what she found. Moreover, when Book repeated the procedure with clinically diagnosed psychopaths from a maximum-security prison, she found something else. The high-scoring "psychopathic" undergraduates in the first study might've been good at identifying weakness, But the *clinical* psychopaths went one better. They explicitly stated it was because of the way people walked. They, like Bundy, knew precisely what they were looking for.

The Men Who Stare at Coats

Angela Book's findings are no flash in the pan. Hers is one of a growing number of studies that have, in recent years, begun to show the psychopath in a new, more complex light: a light somewhat different from the lurid shadows cast by newspaper headlines and Hollywood scriptwriters. The news is difficult to swallow. And it goes down the same way here, in this murderous little corner of Florence, as it does nearly everywhere else: with a healthy dose of skepticism.

"Do you mean," asks Rossi, incredulous, "that there are times when it isn't necessarily a bad thing to be a psychopath?"

"Not only that," I nod, "but there are times when it's actually a good thing—when, by being a psychopath, you in fact have an advantage over other people."

Rossi seems far from convinced, And looking around, it's easy to understand why. Bundy and Gacy aren't exactly the best crowd to fall in with. And, let's face it, when you've got several dozen others knocking about in the wings, it's difficult to see the positives. But the Museum of Serial Killers doesn't tell the full story. In fact, it's not the half of it. As Helen Morrison eloquently elucidated, the fate of a psychopath depends on a whole range of factors, including genes, family background, education, intelligence, and opportunity—and on how they interact.

Jim Kouri, vice president of the U.S. National Association of Chiefs of Police, makes a similar point. Traits that are common among psychopathic serial killers, Kouri observes—a grandiose sense of self-worth, persuasiveness, superficial charm, ruthlessness, lack of remorse, and the manipulation of others—are also shared by politicians and world leaders: individuals running not from the police, but for office. Such a profile, notes Kouri, allows those who present with it to do what they like when they like, completely unfazed by the social, moral, or legal consequences of their actions.

If you are born under the right star, for example, and have as much power over the human mind as the moon has over the sea, you might order the genocide of 100,000 Kurds and shuffle to the gallows with such arcane recalcitrance as to elicit, from even your harshest detractors, perverse, unspoken deference.

"Do not be afraid, doctor," said Saddam Hussein on the scaffold, moments before his execution. "This is for men."

If you are violent and cunning, like real-life "Hannibal Lecter" Robert Maudsley, you might lure a fellow inmate to your cell, smash in his skull with a claw hammer, and sample his brains with a spoon as nonchalantly as if you were downing a soft-boiled egg. (Maudsley, by the way, has been cooped up in solitary confinement for the past thirty years, in a bulletproof cage in the basement of Wakefield Prison in England.)

Or if you are a brilliant neurosurgeon, ruthlessly cool and focused under pressure, you might, like the man I'll call Dr. Geraghty, try your luck on a completely different playing field: at the remote outposts of twenty-first-century medicine, where risk blows in on hundred-mile-an-hour winds and the oxygen of deliberation is thin:

"I have no compassion for those whom I operate on," he told me. "That is a luxury I simply cannot afford. In the theater I am reborn: as a cold, heartless machine, totally at one with scalpel, drill and saw. When you're cutting loose and cheating death high above the snow-line of the brain, feelings aren't fit for purpose. Emotion is entropy, and seriously bad for business. I've hunted it down to extinction over the years."

Geraghty is one of the U.K.'s top neurosurgeons—and though on one level his words send a chill down the spine, on another they make perfect sense. Deep in the ghettos of some of the brain's most dangerous neighborhoods, the psychopath is glimpsed as a lone and ruthless predator, a solitary species of transient, deadly allure. No sooner is the word out than images of serial killers, rapists, and mad, reclusive bombers come stalking down the sidewalks of our minds.

But what if I was to paint you a different picture? What if I was to tell you that the arsonist who burns your house down might also, in a parallel universe, be the hero most likely to brave the flaming timbers of a crumbling, blazing building to seek out, and drag out, your loved ones? Or that the kid with a knife in the shadows at the back of the movie theater might well, in years to come, be wielding a rather different kind of knife at the back of a rather different kind of theater?

Claims like these are admittedly hard to believe. But they're true. Psychopaths are fearless, confident, charismatic, ruthless, and focused. Yet contrary to popular belief, they are not necessarily violent. And if that sounds good, well, it is. Or rather, it can be. It depends, as we've just seen, on what else you've got lurking on the shelves of your personality cupboard. Far from its being an open-and-shut case—you're either a psychopath or you're not—there are, instead, inner and outer zones of the disorder, a bit like the fare zones on a subway map. As we shall see in chapter 2, there is a spectrum of psychopathy along which each of us has our place, with only a small minority of A-listers resident in the "inner city."

One individual, for example, may be ice-cold under pressure, and display about as much empathy as an avalanche (we'll be meeting some like this on the trading floor later), and yet at the same time act neither violently, nor antisocially, nor without conscience. Scoring high on two psychopathic attributes, such an individual may rightly be considered further along the psychopathic spectrum than someone scoring lower on that dyad of traits, yet still not be anywhere near the Chianti-swilling danger zone of a person scoring high on all of them.

Just as there's no official dividing line between someone who plays recreational golf on the weekends and, say, Tiger Woods, so the boundary between a world-class, "hole-in-one" superpsychopath and one who merely "psychopathizes" is similarly blurred. Think of psychopathic traits as the dials and sliders on a studio mixing desk. If you push all of them to max, you'll have a sound track that's no use to anyone. But if the sound track is graded and some controls are turned up higher than others—such as fearlessness, focus, lack of empathy, and mental toughness, for example—you may well have a surgeon who's a cut above the rest.

Of course, surgery is just one instance where psychopathic "talent" may prove beneficial. There are others. Take law enforcement, for example. In 2009, shortly after Angela Book published the results of her study, I decided to perform my own take on it. If, as she'd found, psychopaths really were better at decoding vulnerability, then there had to be applications. There had to be ways in which, rather than being a drain on society, this talent conferred some advantage. Enlightenment dawned when I met a friend at the airport. We all get a bit paranoid going through customs, I mused—even when we're perfectly innocent. But imagine what it would feel like if we *did* have something to hide.

Thirty undergraduate students took part in my experiment, half of whom had scored high on the Self-Report Psychopathy Scale, the other half low. There were also five "associates." The students' job was easy. They had to sit in a classroom and observe the associates' movements as they entered through one door and exited through another, traversing, en route, a small, elevated stage. But there was a catch. The students also had to note who was "guilty": which of the five was concealing a scarlet handkerchief.

To raise the stakes and give the students something to go on, the "guilty" associate was handed £100. If the jury correctly identified the guilty party—if, when the votes were counted, the person with the handkerchief came out on top—then they had to give the money back. If, on the other hand, they got away with it and the finger of suspicion fell more heavily on one of the others, then the "guilty" associate would stand to be rewarded. They would keep the £100.

The nerves were certainly jangling when the associates made their entrance. But which of the students would make the better "customs officers"? Would the psychopaths' predatory instincts prove reliable? Or would their nose for vulnerability let them down?

The results were extraordinary. Over 70 percent of those who scored high on the Self-Report Psychopathy Scale correctly picked out the handkerchief-smuggling associate, compared to just 30 percent of the low scorers. Zeroing in on weakness may well be part of a serial killer's toolkit. But it may also come in handy at the airport.

Psychopath Radar

In 2003, Reid Meloy, professor of psychiatry at the University of California, San Diego, School of Medicine, conducted an experiment that looked at the flip side of the scarlet-handkerchief equation. Sure, traditional "hole-in-one" psychopaths may well have a reputation for sniffing out vulnerability. But they're also known for giving us the creeps. Tales from clinical practice and reports from everyday life are replete with utterances from those who've encountered these ruthless social predators: mysterious, visceral aphorisms such as "the hair stood up on the back of my neck" or "he made my skin crawl." But is there really anything to it? Do our instincts stand up to scrutiny? Are we as good at picking up on psychopaths as psychopaths are at picking up on us?

To find out, Meloy asked 450 criminal justice and mental health professionals whether they'd ever experienced such odd physical reactions when interviewing a psychopathic subject: violent criminals with all the dials on the mixing desk cranked right up to max. The results left nothing to the imagination. Over three-quarters of them said that they had, with female respondents reporting a higher incidence of the phenomenon than males (84 percent compared to 71 percent), and master's/bachelor level clinicians reporting a higher incidence than either those at doctoral level or, on the other side of the professional divide, law enforcement agents (84 percent, 78 percent, and 61 percent, respectively). Examples included "felt like I might be lunch"; "disgust . . . repulsion . . . fascination"; and "an evil essence passed through me."

But what are we picking up on, exactly?

To answer this question, Meloy goes back in time to prehistory and the shadowy, spectral dictates of human evolution. There are a number of theories about how psychopathy might first have developed, and we'll be looking at those a little later on. But an overarching question in the grand etiological scheme of things is from which ontological perspective the condition should actually be viewed: from a clinical standpoint, as a disorder of personality? Or from a game theory standpoint, as a legitimate biological gambit—a life history strategy conferring significant reproductive advantages in the primeval ancestral environment?

Kent Bailey, emeritus professor in clinical psychology at Virginia Commonwealth University, argues in favor of the latter, and advances the theory that violent competition within and between proximal ancestral groups was the primary evolutionary precursor of psychopathy (or, as he puts it, the mind-set of the "warrior hawk").

"Some degree of predatory violence," proposes Bailey, "was required in the seek and kill aspects of hunting large game animals"—and an elite contingent of ruthless "warrior hawks" would presumably have come in handy not only for tracking and killing prey, but also for repelling invasion by similar contingents from other, neighboring groups.

The problem, of course, was how to trust them in peacetime.

Robin Dunbar, professor of evolutionary anthropology at Oxford University, lends support to Bailey's claims. Going back to the time of the Norsemen, between the ninth and twelfth centuries, Dunbar has cited the berserkers as a case in point: the feted Viking warriors who, as the sagas and poems and historical records attest, appear to have fought in a brutal, trance-like fury. But dig a little deeper into the literature and a more sinister picture emerges: of a dangerous elite who could turn against members of the community they were charged to protect, committing savage acts of violence against their countrymen.

Here, proposes Meloy, lies the solution to the mystery: to the prickle at the back of the neck and the long-range evolutionary thinking behind our indwelling "psychopath radar." For if, as Kent Bailey argues, such predatory ancestral individuals were indeed psychopathic, it would follow, from what we know of natural selection, that it wouldn't be a one-way street. More peaceable members of both the immediate and wider communities would, in all probability, themselves evolve a mechanism, the covert neural surveillance technology, to flag and signify danger when entering their cognitive airspace—a clandestine early-warning system that would enable them to beat a retreat.

In the light of Angela Book's work with attack victims and my own investigations into scarlet-handkerchief smuggling, such a mechanism could quite plausibly explain the gender differences revealed by Meloy's experiment. Given psychopaths' enhanced reputation as diabolical emotional sommeliers, their specialized nose for the inscrutable bass notes of weakness, it isn't beyond the bounds of possibility that women, by way of a sneaky Darwinian recompense for greater physical vulnerability, exhibit more intense and more frequent reactions in their presence—as, for exactly the same reason, did the lower-status mental health workers. It's certainly a working hypothesis. The more threatened you feel, the more at risk you are for a break-in, the more important it is to tighten up on security.

Of course, that there existed, in the penumbral days of our ancestors, ruthless, remorseless hunters brutally accomplished in the dark arts of predation is beyond doubt. But that such hunters, with their capacity to second-guess nature, were psychopaths as we know them today is a little more open to question. The stumbling block, diagnostically, is empathy.

In ancestral times, the most prolific and accomplished hunters were not, as one might expect, the most bloodthirsty and indefatigable. They were, in contrast, the most cool and empathetic. They were the ones who were able to assimilate their quarry's mind-set—to see through the eyes of their prey and thus reliably predict its deft, innate trajectories of evasion, its routes and machinations of escape.

To understand why, one need only observe a toddler learning to walk. The gradual development of upright locomotion, of an increasingly bipedal stance, both heralded and facilitated a brand-new era of early hominid grocery shopping. A vertical stance prefigured streamlined, more efficient mobility, enabling our forebears on the African savannah to forage and hunt for considerably longer periods than quadrupedal locomotion would have allowed.

But "persistence hunting," as it's known in anthropology, has problems of its own. Wildebeest and antelopes can easily outsprint a human. They can vanish over the horizon. If you can accurately predict where they might eventually stop—either by looking for clues that they've left behind in their flight or by reading their minds, or both—you can marginally increase your chances of survival.

So if predators demonstrate empathy, and in some cases even enhanced empathy, how can they really be psychopaths? If there's one thing most people agree on, it's that psychopaths exhibit a marked absence of feeling, a singular lack of understanding of others. How do we square the circle? Help is at hand in the form of cognitive neuroscience, with a bit of an assist from some fiendish moral philosophy.

Trolleyology

Joshua Greene, a psychologist, neuroscientist, and philosopher at Harvard University, has observed how psychopaths unscramble moral dilemmas, how their brains respond inside different ethical compression chambers. As I described in my previous book, *Split-Second Persuasion*, he's stumbled upon something interesting. Far from being uniform, empathy is schizophrenic. There are two distinct varieties: hot and cold.

Consider, for example, the following conundrum (case 1), first proposed by the philosopher Philippa Foot:

A railway trolley is hurtling down a track. In its path are five people who are trapped on the line and cannot escape. Fortunately, you can flip a switch that will divert the trolley down a fork in the track away from the five people—but at a price. There is another person trapped down that fork, and the trolley will kill them instead. Should you hit the switch?

Most of us experience little difficulty when deciding what to do in this situation. Though the prospect of flipping the switch isn't exactly a nice one, the utilitarian option—killing just one person instead of five—represents the "least worst choice." Right?

Now consider the following variation (case 2), proposed by the philosopher Judith Jarvis Thomson:

As before, a railway trolley is speeding out of control down a track toward five people. But this time, you are standing behind a very large stranger on a footbridge above the tracks. The only way to save the five people is to heave the stranger over. He will fall to a certain death. But his considerable girth will block the trolley, saving five lives. Should you push him?

Here, you might say we're faced with a "real" dilemma. Although the score in lives is precisely the same as in the first example (five to one), playing the game makes us a little more circumspect and jittery. But why? Greene believes he has the answer—and that it's got to do with different climatic regions in the brain.

Case 1, he proposes, is what we might call an impersonal moral dilemma. It involves those areas of the brain, the prefrontal cortex and posterior parietal cortex (in particular, the anterior paracingulate cortex, the temporal pole, and the superior temporal sulcus), principally implicated in our objective experience of cold empathy: in reasoning and rational thought.

Case 2, on the other hand, is what we might call a personal moral dilemma. It hammers on the door of the brain's emotion center, known as the amygdala—the circuit of hot empathy.

Just like most normal members of the population, psychopaths make pretty short work of the dilemma presented in case 1. They flip the switch, and the train branches accordingly, killing just the one person instead of five. However—and this is where the plot thickens—quite unlike normal people, they also make pretty short work of case 2. Psychopaths, without batting an eye, are perfectly happy to chuck the fat guy over the side, if that's how the cookie crumbles.

To compound matters further, this difference in behavior is mirrored, rather distinctly, in the brain. The pattern of neural activation in both psychopaths and normal people is pretty well matched on the presentation of impersonal moral dilemmas—but dramatically diverges when things start to get a bit more personal.

Imagine that I was to pop you into an fMRI machine* and then

*In fMRI, or functional magnetic resonance imaging, a giant magnet surrounds the subject's head. Changes in the direction of the magnetic field induce hydrogen

present you with the two dilemmas. What would I observe as you went about negotiating their mischievous moral minefields? Well, around the time that the nature of the dilemma crossed the border from impersonal to personal, I would see your amygdala and related brain circuits—your medial orbitofrontal cortex, for example—light up like a pinball machine. I would witness the moment, in other words, when emotion puts its money in the slot.

But in a psychopath, I would see only darkness. The cavernous neural casino would be boarded up and derelict. And the crossing from impersonal to personal would pass without any incident.

This distinction between hot and cold empathy, the kind of empathy that we "feel" when observing others, and the steely emotional calculus that allows us to gauge, coolly and dispassionately, what another person might be thinking, should be music to the ears of theorists such as Reid Meloy and Kent Bailey. Sure, psychopaths may well be deficient in the former variety, the touchy-feely type. But when it comes to the latter commodity, the kind that codes for "understanding" rather than "feeling"; the kind that enables abstract, nerveless prediction, as opposed to personal identification; the kind that relies on symbolic processing instead of affective symbiosis—the cognitive skill set possessed by expert hunters and cold readers, not just in the natural environment, but in the human arena, too—then psychopaths are in a league of their own. They fly even better on one empathy engine than on two—which is, of course, just one of the reasons why they make such good persuaders. If you know where the buttons are and don't feel the heat when you push them, then chances are you're going to hit the jackpot.

The empathy divide is certainly music to the ears of Robin Dunbar, who, when he's not reading up on berserkers, can sometimes be found in the Magdalen College Senior Common Room. One afternoon, over tea and cakes in an oak-paneled alcove overlooking the cloisters, I tell him about the railway trolleys and the difference they reveal be-

---

atoms in the brain to emit radio signals. These signals increase when the level of blood oxygen goes up, indicating which parts of the brain are most active.

tween psychopathic and normal brain function. He's not in the least bit surprised.

"The Vikings had a pretty good run of things back in their day," he points out. "And the berserkers certainly didn't do anything to dispel their reputation as a people not to be messed with. But that was their job. Their role was to be more ruthless, more cold-blooded, more savage than the average Viking soldier, because . . . that was exactly who they were! They *were* more ruthless, more cold-blooded, more savage than the average Viking soldier. If you'd wired up a berserker to a brain scanner and presented him with the trolley dilemma, I'm fairly certain I know what you'd have got. Exactly what you get with psychopaths. Nothing. And the fat bloke would've been history!"

I butter myself a scone.

"I think every society needs particular individuals to do its dirty work for it," he continues. "Someone who isn't afraid to make tough decisions. Ask uncomfortable questions. Put themselves on the line. And a lot of the time those individuals, by the very nature of the work that they're tasked to do, aren't necessarily going to be the kind of people who you'd want to sit down and have afternoon tea with. Cucumber sandwich?"

Daniel Bartels at Columbia University and David Pizarro at Cornell couldn't agree more—and they've got documentary evidence to prove it. Studies have shown that approximately 90 percent of people would refuse to push the stranger off the bridge, even though they know that if they could just overcome their natural moral squeamishness, the body count would be one-fifth as high. That, of course, leaves 10 percent unaccounted for: a less morally hygienic minority who, when push quite literally comes to shove, have little or no compunction about holding another person's life in the balance. But who is this unscrupulous minority? Who is this 10 percent?

To find out, Bartels and Pizarro presented the trolley problem to more than two hundred students, and got them to indicate on a four-point scale how much they were in favor of shoving the fat guy over the side—how "utilitarian" they were. Then, alongside the trolleyological question, the students also responded to a series of personality items specifically designed to measure resting psychopathy levels. These included statements such as "I like to see fistfights" and "The best way to handle people is to tell them what they want to hear" (agree/disagree on a scale of one to ten).

Could the two constructs—psychopathy and utilitarianism—possibly be linked? Bartels and Pizarro wondered. The answer was a resounding yes. Their analysis revealed a significant correlation between a utilitarian approach to the trolley problem (push the fat guy off the bridge) and a predominantly psychopathic personality style. Which, as far as Robin Dunbar's prediction goes, is pretty much on the money. But which, as far as the traditional take on utilitarianism goes, is somewhat problematic. In the grand scheme of things, Jeremy Bentham and John Stuart Mill, the two nineteenth-century British philosophers credited with formalizing the theory of utilitarianism, are generally thought of as good guys.

"The greatest happiness of the greatest number is the foundation of morals and legislation," Bentham once famously articulated.

Yet dig a little deeper and a trickier, quirkier, murkier picture emerges—one of ruthless selectivity and treacherous moral riptides. Crafting that legislation, for example, excavating those morals, will inevitably necessitate riding roughshod over someone else's interests. Some group or cause, through the simple lottery of numbers, has to bite the bullet for the sake of the "greater good." But who has got the balls to pull the trigger? Bartels and Pizarro may well have found a pattern in the lab. But what about in everyday life? Is this where the psychopath really comes into his own?

Dark Side of the Moon Landing

The question of what it takes to succeed in a given profession, to deliver the goods and get the job done, is not all that difficult when it comes down to it. Alongside the dedicated skill set necessary to perform one's specific duties, there exists, in law, in business, in whatever field of endeavor you care to mention, a selection of traits that code for high achievement.

In 2005, Belinda Board and Katarina Fritzon of the University of Surrey conducted a survey to find out precisely what it was that made business leaders tick. What, they wanted to know, were the key facets of personality that separated those who turn left when boarding an airplane from those who turn right?

Board and Fritzon took three groups—business managers, psychiatric patients, and hospitalized criminals (both those who were psychopathic and those suffering from other psychiatric illnesses)—and compared how they fared on a psychological profiling test.

Their analysis revealed that a number of psychopathic attributes were actually more common in business leaders than in so-called disturbed criminals—attributes such as superficial charm, egocentricity, persuasiveness, lack of empathy, independence, and focus—and that the main difference between the groups was in the more "antisocial" aspects of the syndrome: the criminals' lawbreaking, physical aggression, and impulsivity dials (to return to our analogy of earlier) were cranked up higher.

Other studies seem to confirm the "mixing desk" picture: the borderline between functional and dysfunctional psychopathy depends not on the presence of psychopathic attributes per se, but rather on their levels and the way they're combined. Mehmet Mahmut and his colleagues at Macquarie University have recently shown that patterns of brain dysfunction (specifically, in relation to the orbital frontal cortex, the area of the brain that regulates the input of the emotions in decision making) observed in both criminal and noncriminal psychopaths exhibit dimensional rather than discrete differences. This, he suggests, means that the two groups should not be viewed as qualitatively distinct populations, but rather as occupying different positions on the same neuropsychological continuum.

In a similar (if less high-tech) vein, I asked a class of first-year undergraduates to imagine they were managers in a job placement company. "Ruthless, fearless, charming, amoral, and focused," I told them. "Suppose you had a client with that kind of profile. To which line of work do you think they might be suited?"

Their answers, as we shall see a little later on in the book, couldn't have been more insightful. CEO, spy, surgeon, politician, the military . . . they all popped up in the mix. Along with serial killer, assassin, and bank robber.

"Intellectual ability on its own is just an elegant way of finishing second," one successful CEO told me. "Remember, they don't call it a greasy pole for nothing. The road to the top is hard. But it's easier to climb if you lever yourself up on others. Easier still if they think something's in it for them."

Jon Moulton, one of London's most successful venture capitalists, agrees. In a recent interview with *The Financial Times*, he lists determination, curiosity, and insensitivity as his three most valuable character traits.

No prizes for guessing the first two. But insensitivity? "The great thing about insensitivity," explains Moulton, "is that it lets you sleep when others can't."

If the idea of psychopathic traits lending a hand in business doesn't come as too great a surprise, then how about in space? Blasting psychopaths off deep into the cosmos does not, I am sure, given their terrestrial reputation, particularly inspire confidence—and psychopathic qualities, you'd think, might not exactly be foremost among NASA's prohibitively exclusive selection criteria for astronauts. But there's a story I once heard that provides a graphic illustration of how the refrigerated neurology that showed up on Robert Hare's brain scans can, in certain situations, confer real benefits; how the reptilian focus and crystalline detachment of neurosurgeon James Geraghty can sometimes code for greatness not just in the boardroom, the courtroom, and the operating theater. But in another world entirely.

The story goes like this. On July 20, 1969, as Neil Armstrong and his partner Buzz Aldrin zipped across the lunar landscape looking for a place to set their module down, they came within seconds of crash-landing. The problem was geology. There was just too much of it. And fuel: too little. Rocks and boulders lay scattered all over the place, making a safe approach impossible. Aldrin mopped his brow. With one eye on the gas gauge and the other on the terrain, he issued Armstrong a stark ultimatum: Get this thing down—and fast!

Armstrong, however, was decidedly more phlegmatic. Maybe—who knows?—he'd never had time for twitchy backseat drivers. But with the clock running down, the fuel running out, and the prospect of death by gravity an ever-increasing possibility, he coolly came up with a game plan. Aldrin, he instructed, was to convert into seconds the amount of fuel they had left, and to start counting down. Out loud.

Aldrin did as he was asked.

Seventy . . . sixty . . . fifty . . .

As he counted, Armstrong scrutinized the moon's unyielding topography.

Forty . . . thirty . . . twenty . . .

Still the landscape refused to give an inch.

Then, with just ten seconds remaining, Armstrong spotted his chance: a silver oasis of nothingness just below the horizon. Suddenly, imperceptibly, like a predator closing in on its prey, his brain narrowed its focus. As if he were on a practice run, he maneuvered the craft deftly toward the drop zone and performed, in the only clearing for miles, the perfect textbook touchdown. One giant leap for mankind. But almost, very nearly, one giant cosmological screw-up.

Bomb-Disposal Experts—What Makes Them Tick?

This extraordinary account of incredible interplanetary insouciance epitomizes life on the horizons of possibility, where triumph and disaster share a fraught and fragile frontier and cross-border traffic flows freely. This time, however, the road to disaster was closed. And Neil Armstrong's coolness under fire rescued from cosmological calamity one of the greatest feats ever in the history of human achievement. But there's more. His heart rate, reports revealed later, barely broke a sweat. He might as well have been landing a job in a gas station as a spaceship on the moon. Some freakish strain of cardiovascular genius? The science suggests not.

Back in the 1980s, Harvard researcher Stanley Rachman found something similar with bomb-disposal operatives. What, Rachman wanted to know, separated the men from the boys in this high-risk, high-wire profession? All bomb-disposal operatives are good. Otherwise they'd be dead. But what did the stars have that the lesser luminaries didn't?

To find out, he took a bunch of experienced bomb-disposal operatives—those with ten years or more in the business—and split them into two groups: those who'd been decorated for their work, and those who hadn't. Then he compared their heart rates in the field on jobs that demanded particularly high levels of concentration.

What he turned up was astonishing. Whereas the heart rates of all the operatives remained stable, something quite incredible happened with the ones who'd been decorated. Their heart rates actually went down. As soon as they entered the danger zone (or the "launch pad," as one guy I spoke with put it), they assumed a state of cold, meditative focus: a mezzanine level of consciousness in which they became one with the device they were working on.

Follow-up analysis probed deeper, and revealed the cause of the disparity: confidence. The operatives who'd been decorated scored higher on tests of core self-belief than their non-decorated colleagues.

It was conviction that made them tick.

Stanley Rachman knows all about the fearless arctic neurology of the psychopath. And his findings were certainly explosive. So much so that he raised the question himself: Should we be keeping a closer eye on our bomb-disposal operatives? His conclusion seems pretty clear: "The operators who received awards for courageous/fearless behavior," he reports, "were free of psychological abnormalities or antisocial behavior." In contrast, he points out, "most descriptions of psychopathy include adjectives such as 'irresponsible' and 'impulsive'"—adjectives that, in his experience, did not befit any of his case studies.

In the light of Belinda Board and Katarina Fritzon's 2005 survey, however—which, if you will recall, demonstrated that a number of psychopathic traits were more prevalent among business leaders than among diagnosed criminal psychopaths—Rachman's comments beg the question of what, precisely, we mean when we use the word "psychopath." Not all psychopaths are as wholly undomesticated, as socially feral, as he might have us believe. In fact, the standout implication of Board and Fritzon's study is the suggestion that it is precisely this "antisocial" wing of the disorder, comprising the elements of impulsivity and irresponsibility, that either "makes or breaks" psychopaths—that codes them, depending on how high these particular personality dials are turned up, for dysfunction or success.

To throw another methodological monkey wrench in the works, it turns out that bomb-disposal operatives aren't the only ones who experience a drop in heart rate when they get down to business. Relationship experts Neil Jacobson and John Gottman, authors of the popular book *When Men Batter Women*, have observed identical cardiovascular profiles in certain types of abusers, who, research has shown, actually become more relaxed when beating up their partners than when they're lounging in an armchair with their eyes closed.

In their widely cited typology of abusers, Jacobson and Gottman refer to individuals with this type of profile as "Cobras." Cobras, unlike their opposite numbers, the "Pit Bulls," attack swiftly and ferociously, and remain in control. They possess a grandiose sense of entitlement to whatever they feel like, whenever they feel like it. In addition, as their name suggests, they become calm and focused prior to launching an offensive. Pit Bulls, on the other hand, are emotionally more volatile, more prone to let things fester—and then fly off the handle. Further comparisons between these two groups make interesting reading:

COBRA	PIT BULL
Displays violence toward others	Usually only violent toward partner
Feels little remorse	Shows some level of guilt
Motivated by the desire for immediate gratification	Motivated by fear of abandonment
Able to let go and move on	Obsessive; often stalks victim
Feels superior	Adopts the role of "victim"
Fast talker; able to spin a story to the authorities	Greater emotionally lability
Charming and charismatic	Depressed and introverted
Control means not being told what to do	Control means constant monitoring of partner
Traumatic upbringing; violence prevalent in family	Some degree of violence in family background
Impermeable to therapeutic intervention	Sometimes benefits from treatment programs

Devastating fearlessness may well be descended from courage, as Rachman proposes in bomb disposal. It may well habituate through repeated exposure to danger. But there are some individuals who claim it as their birthright, and whose basic biology is so fundamentally different from the rest of ours as to remain, both consciously and unconsciously, completely impermeable to even the minutest trace of anxiety antigens. I know, because I've tested them.

The Scent of Fear

If you've ever been spooked by inflight turbulence, or become slightly uneasy when a train has stopped in a tunnel, or simply experienced that indefinable feeling of dread that "something just isn't quite right," you may have been responding to the fears of those around you just as much as to anything else. In 2009, Lilianne Mujica-Parodi, a cognitive neuroscientist at Stony Brook University in New York, collected sweat from the armpits of first-time skydivers as they hurtled toward the ground at terminal velocity. Back in the lab, she then transferred the sweat—from absorbent pads secured under volunteers' arms—as well as samples of normal "fear-lite" treadmill sweat to a specially calibrated nebulizer box, and waved it under the noses of a second bunch of volunteers as they sat in an fMRI scanner.

Guess what? Even though none of the volunteers had any idea what they were inhaling, those who were exposed to the fear sweat showed considerably more activity in their brains' fear-processing zip codes (their amygdalae and hypothalami) than those who'd breathed the exercise sweat. In addition, on an emotion recognition task, volunteers who had inhaled the fear sweat were 43 percent more accurate at judging whether a face bore a threatening or neutral expression than those who'd gotten the workout sweat.

All of which raises a rather interesting question: Can we "catch" fear in the same way we catch a cold? Mujica-Parodi and her team certainly seem to think so. In the light of their findings, they allude to the possibility that "there may be a hidden biological component to human social dynamics, in which emotional stress is, quite literally, 'contagious.' "

Which raises, of course, an even more interesting question: What about immunity? Are some of us more likely to come down with the fear bug than others? Do some of us have more of a "nose" for it?

To find out, I ran a variation on Mujica-Parodi's study. First, I showed one group of volunteers a scary movie (*Candyman*) and got a second group on a treadmill. Next, I collected their sweat. Third, I bottled it (so to speak). Finally, I squirted it up the noses of a second group of volunteers as they played a simulated gambling game.

The game in question was the Cambridge Gamble Task, a computerized test of decision making under risk. The test comprises a sequence of trials in which participants are presented with an array of ten boxes (either red or blue in color) and must guess, on each trial, which of those boxes conceals a yellow token. The proportion of colored boxes varies from trial to trial (e.g., 6 red and 4 blue; 1 blue and

9 red), and participants start off with a total of 100 points—a fixed proportion of which (5, 25, 50, 75, or 95 percent) they must bet on the outcome of the first trial. What happens then is contingent on the result. Depending on whether they win or lose, the amount wagered is either added to or subtracted from their initial tally, and the protocol is repeated, with a rolling total, on all subsequent trials. Higher bets are associated with higher risk.

If Mujica-Parodi's theory held any water, then the prediction was pretty straightforward. Volunteers who inhaled the *Candyman* sweat would exercise greater caution and gamble more conservatively than those who inhaled the treadmill sweat.

But there was a catch. Half the volunteers were psychopaths. Would the psychopaths, noted for their coolness under pressure, be immune to others' stress? Like expert hunters and trackers, might they be hypervigilant for visual cues of vulnerability—as Angela Book discovered—yet chemically impervious to olfactory ones?

The results of the experiment couldn't have been any clearer. Exactly as predicted by Mujica-Parodi's findings, the non-psychopathic volunteers played their cards pretty close to their chests when exposed to the fear sweat, staking lower percentages on outcomes. But the psychopaths remained unfazed. Not only were they more daring to start with, they were also more daring to finish with, continuing to take risks even when pumped full of "fear." Their neurological immune systems seemed to immediately crack down on the "virus," adopting a zero tolerance policy on anxiety, while the rest of us just allow it to spread.

Double-Edged Sword

Glimpsed in passing through a shop window, or more likely, these days, on Amazon, "The Wisdom of Psychopaths" may seem rather an odd conglomeration of words to appear on the front cover of a book. Eye-catching, maybe. But odd, most certainly. The jarring juxtaposition of those two existential monoliths, "wisdom" and "psychopaths," precipitates, one would have thought, little semantic compromise, little in the way of constructive, meaningful dialogue around the logic-scored scientific negotiating table.

And yet the core, underlying thesis that psychopaths are in possession of wisdom is a serious one. Not, perhaps, wisdom in the traditional sense of that word: as an emergent property of advancing years and cumulative life experience. But as an innate, ineffable function of their being.

Consider, for example, the following analogy from someone we'll be meeting later.

A psychopath.

Within, I should add, the rarefied, cloistral confines of a maximum-security personality disorder unit:

"A powerful top-of-the-range sports car is neither a good thing nor a bad thing in and of itself, but depends on the person who's sitting behind the wheel. It may, for instance, permit a skilled and experienced motorist to get his wife to the hospital in time to give birth to their child. Or, in a parallel universe, run an eighteen-year-old and his girlfriend off a cliff.

"In essence, it's all in the handling. Quite simply, the skill of the driver . . ."

He's right. Perhaps the one stand-alone feature of the psychopath, the ultimate "killer" difference that distinguishes the psychopathic personality from the personalities of most "normal" members of the population, is that psychopaths don't give a damn what their fellow citizens think of them. They simply couldn't care less how society, as a whole, might contemplate their actions. In a world in which image and branding and personal reputation are more sacrosanct than ever—what are we up to now: 500 million on Facebook? 200 million videos on YouTube? One closed-circuit TV camera for every 20 people in the U.K.?—this constitutes, no doubt, one of the fundamental reasons why they run into so much trouble.

And, of course, why we find them so beguiling.

Yet it may also predispose to heroism and mental toughness, to estimable qualities such as courage, integrity, and virtue: the ability, for instance, to dart into blazing buildings to save the lives of those inside. Or to push fat guys off bridges and stop runaway trains in their tracks.

Psychopathy really is like a high performance sports car. It's a double-edged sword that inevitably cuts both ways.

In the following chapters, I'll chronicle, in scientific, sociological, and philosophical detail, the story of this double-edged sword and the unique psychological profile of the individuals that wield it. We begin by looking at who, precisely, the psychopath really is (if not the monster we usually think of). We travel through both the inner and outer zones of the psychopathic metropolis, cruising the ultraviolent downtown ghettos and the lighter, leafier, more visitor-friendly suburbs.

As on any scale or spectrum, both ends have their poster-boy Hall of Famers. At one end we have the Dahmers and Lecters and Bundys—the Rippers and Slashers and Stranglers. At the other extreme we have the antipsychopaths: elite spiritual athletes like Tibetan Buddhist monks, who, through years of black-belt meditation in remote Himalayan monasteries, feel nothing but compassion. In fact, the latest research from the field of cognitive neuroscience suggests that the spectrum might be circular . . . that across the neural dateline of sanity and madness, the psychopaths and antipsychopaths sit within touching distance of each other. So near and yet so far.

From secluded neural datelines, we'll shift our focus to cognitive archaeology, and having sketched out the coordinates of modern-day psychopathy, we'll go in search of its origins. Using the instruments of game theory and cutting-edge evolutionary psychology, we will reconstruct the conditions, deep in our ancestral past, under which psychopaths might have evolved. And we'll explore the possibility—as profound as it is disturbing—that in twenty-first-century society they're continuing to evolve, and that the disorder is becoming adaptive.

We'll consider, in depth, the advantages of being a psychopath—or rather, in some situations at least, having those dials turned up a little higher than normal. We'll look at the fearlessness. The ruthlessness. The "presence" (psychopaths tend to blink just a little bit less than the rest of us, a physiological aberration that often helps give them their unnerving, hypnotic air).* Devastating, dazzling, and super-confident are the epithets that one often hears about them. Not, as one might expect, from themselves. But from their victims! The irony is plain as day. Psychopaths appear, through some Darwinian practical joke, to possess the very personality characteristics that many of us would die for. Indeed, that many *have* died for—the reason, of course, why our old friend Fabrizio Rossi had trouble believing that anything good could possibly come out of the crawl space.

We'll go behind the scenes of one of the most feted psychopath units in the world and get a psychopath's take on the problems, dilemmas, and challenges that each of us faces during the course of everyday life. We'll catch up with the neuroscientist and psychopath hunter Kent Kiehl as he trawls an eighteen-wheel truck, housing a custom-built fMRI scanner, around America's state penitentiaries.

And in a groundbreaking, one-off experiment, I finally undergo a "psychopath makeover" myself as a world-renowned expert in transcranial magnetic stimulation simulates, with the aid of some remote, noninvasive neurosurgery, a psychopathic brain state inside my own head (it's worn off).

As *The Wisdom of Psychopaths* unfolds, the truth, like a remorseless predator itself, slowly begins to close in. Sure, these guys might sting us. But they might also save our lives. Either way, they certainly have something to teach us.

*Many people who come into contact with psychopaths subsequently comment on their unusually piercing eyes—a fact not lost on numerous Hollywood screenwriters. The precise reason for this is unclear. On the one hand, blink rate is a reliable index of resting anxiety levels. And so, as mentioned, psychopaths, on average, blink slightly less than the rest of us—an autonomic artifact that may well contribute to their intense "reptilian" aura. On the other hand, however, it's also been speculated that psychopaths' intense gaze may reflect enhanced, predatory concentration levels: like the world's top poker players, they are continually psychologically "frisking" their "opponents" for key emotional tells.

TWO

WILL THE REAL PSYCHOPATH PLEASE STAND UP?

> Who in the rainbow can draw the line where the violet tint ends and the orange tint begins? Distinctly we see the difference of the colors, but where exactly does the one first blendingly enter into the other? So with sanity and insanity.
>
> —HERMAN MELVILLE

Whydunnit?

There's a story making the rounds on the Internet, and it goes like this. While attending her mother's funeral, a woman meets a man she's never seen before. She thinks he's incredible. She believes him to be her soul mate and falls for him instantly. But she never asks for his number, and when the funeral is over, cannot track him down. A few days later she kills her sister. Why?

Take a little time before you answer. Apparently, this simple test can determine whether or not you think like a psychopath. What motive could the woman possibly have for taking her sister's life? Jealousy? She subsequently finds her sister in bed with the man? Revenge? Both plausible. But wrong. The answer, assuming you think like a psychopath, is this: because she was hoping the man would turn up again at her sister's funeral.

If this was the solution that you came up with . . . don't panic. Actually, I lied. Of course it doesn't mean you think like a psychopath. Like a great many things you stumble upon on the Internet, this tale contains about as much truth as Bernie Madoff's profit-and-loss account. Sure, on the face of it, the woman's strategy is certainly psychopathic, there's no disputing that: cold, ruthless, emotionless, and myopically self-interested. But unfortunately, there's a problem. When I gave this test to some real psychopaths—rapists, murderers, pedophiles, and armed robbers—who'd been properly diagnosed using standardized clinical procedures, guess what happened? Not one of them came up with the "follow-up funeral" motive. Instead, nearly all of them came up with the "romantic rivalry" rationale.

"I might be nuts," one of them commented. "But I'm certainly not stupid."

Scott Lilienfeld is professor of psychology at Emory University in Atlanta, and one of the world's leading experts on psychopaths—or, as he puts it, successful psychopaths: those more likely to make a killing on the stock market than down some dark, trash-can-strewn alley. As we tuck into alligator tacos in a deep-fried Southern diner just a mile or two from his office, I ask him about the funeral conundrum. What's going on here? What is it about this kind of stuff that gets us so excited? The question hits a nerve.

"I think the appeal of items like this lies in their cleanliness," he says. "There's something reassuring in the idea that by asking just the one question we can somehow unmask the psychopaths in our midst and be able to protect ourselves from them. Unfortunately, however, it just ain't that simple. Sure, we can work out who they are. But it takes more than just the one question. It takes quite a few of them."

He's right. "Silver bullet" questions, which through some fiendish sleight of mind can somehow reveal our true psychological colors, just don't exist in the real world. Personality is far too complex a construct to give up its secrets purely on the basis of a one-off, one-shot parlor game. In fact, experts in the field have risen to the task of firing quite a few bullets down the years. And it's only relatively recently that they've thought about calling a truce.

The Personality Hunters

Personality has a long history—or rather, measuring it does. It began in ancient Greece, with Hippocrates (460–377 B.C.), the father of Western medicine. Drawing on the wisdom of earlier traditions still (the celestial calculus of Babylonian astrology, for example) that had ghosted across the Levant from the sages of ancient Egypt and the mystics of Mesopotamia, Hippocrates discerned four distinct temperaments in the canon of human emotions: sanguine, choleric, melancholic, and phlegmatic (see figure 2.1).

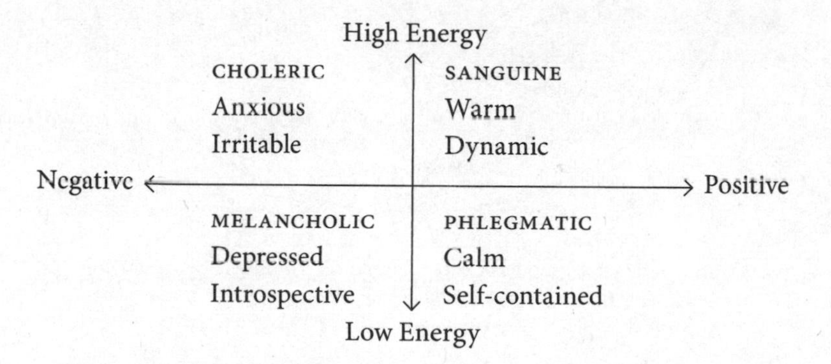

Figure 2.1. Hippocrates' four temperaments

After Hippocrates, not a lot happened for two and a half millennia. Then, in 1952, the British psychologist Hans Eysenck gave the father of Western medicine's ancient dyadic taxonomy a new lease on life. Following exhaustive questionnaire analysis and in-depth clinical interviews, Eysenck proposed that human personality comprised two core dimensions: introversion/extraversion and neuroticism/stability (a third, psychoticism, characterized by aggression, impulsiveness, and egocentricity, was added later). These two dimensions, when set out orthogonally, perfectly encapsulated the four classical temperaments originally identified by Hippocrates (see figure 2.2).

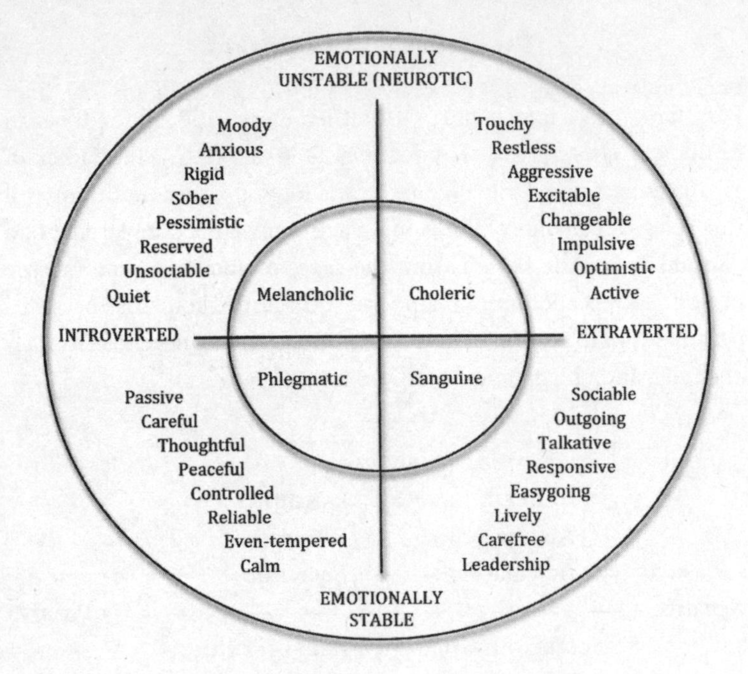

Figure 2.2. Eysenck's model of personality incorporating Hippocrates' four temperaments (from Eysenck and Eysenck, 1958)

The choleric personality (anxious; irritable) mapped onto Eysenck's neurotic extraversion; the melancholic (depressed; introspective) onto neurotic introversion; the sanguine (warm; dynamic) onto emotionally stable extraversion; and the phlegmatic (calm; self-contained) onto emotionally stable introversion. Hippocrates, it appeared, wasn't just the father of Western medicine, but of human nature too.

Eysenck's two-stroke model of personality was positively anorexic compared to the gargantuan corpus of traits unearthed by the American psychologist Gordon Allport some twenty years earlier. In line with the so-called lexical hypothesis of personality, which stipulated that all significant character-related terms would, quite literally by definition, be encoded into language, Allport set sail into the deep, wordy waters of *Webster's New International Dictionary* on a fishing trip. How many personality-related adjectives were out there? he wondered. The answer, it transpired, was quite a few—and he reemerged onto dry land with a haul of nearly 18,000. This list, after junking those descriptions relating to temporary, rather than enduring, traits (e.g., elated, shamefaced), Allport trimmed down to a more tractable 4,500.

But it wasn't until University of Illinois psychologist Raymond Cattell got hold of Allport's list in 1946, at the same time that Eysenck was working on his model, that personality theorists really had something to play with. Eliminating synonyms, and introducing some additional items gleaned from laboratory research, Cattell arrived at a word count of 171. Then he got down to business. Using these descriptions to generate rating scales, he handed them out to volunteers. The task was refreshingly simple: to assess their acquaintances based on the tags provided.

Analysis revealed a galactic personality structure fiendishly composed of thirty-five major trait clusters referred to by Cattell, somewhat esoterically, as the "personality sphere." Over the next decade, further refinement, with the aid of first-generation computers and the embryonic sorcery of factor analysis,* whittled it down even further, to just 16 primary factors. There Cattell called it a day.

LOW SCORERS	FACTOR	HIGH SCORERS
Reserved	Warmth	Outgoing
Less Intelligent	Reasoning	More intelligent
Emotionally Reactive	Emotional Stability	Emotionally Stable
Submissive	Dominance	Assertive
Serious	Liveliness	Happy-Go-Lucky
Nonconforming	Rule Consciousness	Conscientious
Shy	Social Boldness	Uninhibited
Tough-Minded	Sensitivity	Tender-Minded

(continued)

* Factor analysis is a statistical technique used to discover simple patterns in the relationships between different variables. In particular, it seeks to discover whether the observed variables can be explained in terms of a much smaller number of variables, called *factors*. As an example, in Cattell's model, the superordinate factor "warmth" was distilled from constituent descriptors such as "friendly," "empathic," and "welcoming."

LOW SCORERS	FACTOR	HIGH SCORERS
Trusting	Vigilance	Suspicious
Practical	Abstractedness	Imaginative
Forthright	Privateness	Discreet
Self-Assured	Apprehension	Apprehensive
Conservative	Openness to Change	Experimenting
Group-Oriented	Self-Reliance	Self-Sufficient
Unexacting	Perfectionism	Precise
Relaxed	Tension	Tense

Figure 2.3. Cattell's 16 primary personality factors (adapted from Cattell, 1957)

Fortunately for occupational psychologists, however, and those now working in the field of human resources, subsequent theorists pressed on. In 1961, two U.S. Air Force researchers, Ernest Tupes and Raymond Christal, succeeded in distilling Cattell's traits into just five recurring factors. These they labeled Surgency, Agreeableness, Dependability, Emotional Stability, and Culture. More recently, over the last twenty years or so, the work of Paul Costa and Robert McCrae at the National Institutes of Health has led to the development of a standardized test of personality called the NEO Personality Inventory.

Psychologists don't really do consensus if they can help it. But in this case it's hard to avoid. Openness to Experience, Conscientiousness, Extraversion, Agreeableness, and Neuroticism—think OCEAN—comprise the genome of human personality. And we're all the sum of our parts. We are not numbers, as Patrick McGoohan famously asserted in *The Prisoner*. Rather, we are a constellation of numbers. Each of us, in the infinite algorithmic firmament of personality space, has our own unique coordinates depending on precisely where we fall along each of these five dimensions.* Or, as they're commonly referred to, the "Big Five."

* If you want to find out who *you* are from your personality, you may like to try an abbreviated version of the Big Five personality test, which you can find at www.wisdomofpsychopaths.com.

Give Me Five

To the casual observer, of course, personality presents as continuous and uniform. It's only when sifted through a prism of mathematical scrutiny that it formally degrades into its five constituent components. The Big Five, you might say, correspond to those psychologically indivisible "primary colors" of personality, anchored at either endpoint by polar opposite character traits: a spectrum of identity that innervates us all.

These traits, together with a brief description of the set of personal attributes associated with each dimension, are laid out in figure 2.4.

FACTOR	DESCRIPTORS
Openness to Experience	Imaginative . . . . . . Practical Likes Variety . . . . . . Likes Routine Independent . . . . . . Conforming
Conscientiousness	Organized . . . . . . Disorganized Careful . . . . . . Careless Self-Disciplined . . . . . . Impulsive
Extraversion	Sociable . . . . . . Retiring Fun-Loving . . . . . . Sober Affectionate . . . . . . Reserved
Agreeableness	Softhearted . . . . . . Ruthless Trusting . . . . . . Suspicious Helpful . . . . . . Uncooperative
Neuroticism	Worried . . . . . . Calm Insecure . . . . . . Secure Self-Pitying . . . . . . Self-Satisfied

Figure 2.4. The Big Five factor model of personality (McCrae and Costa, 1999, 1990)

Unsurprisingly, perhaps, occupational psychologists have gotten a lot of mileage out of the NEO (and other Big Five personality tests like it). They've handed it out to employees in virtually every profession you can think of to fathom the precise relationship between psychological makeup and success in the workplace. In doing so, they've found a striking connection between temperament and job type. Between how we're wired and where we're hired.

Openness to Experience has been shown to play an important role in professions in which original thought or emotional intelligence is the order of the day—professions such as consultancy, arbitration, and advertising—while individuals scoring lower on this dimension tend to do better in manufacturing or mechanical jobs. Employees scoring medium to high on Conscientiousness (too high and you slip across the border into obsession, compulsion, and perfectionism) tend to excel across the board, the opposite being true for those posting lower scores. Extroverts do well in jobs that require social interaction, while introverts do well in more "solitary" or "reflective" professions, such as graphic design and accountancy. Rather like Conscientiousness, Agreeableness is pretty much a universal facilitator of performance, but shows up particularly prominently in occupations where the emphasis is on teamwork or customer service, like nursing and the armed forces, for instance. But unlike Conscientiousness, having lower levels of this trait can also come in handy—in bruising, cutthroat arenas such as the media, for example, where egos clash and competition for resources (ideas, stories, commissions) is often fierce.

Last, we have Neuroticism, arguably the most precarious of the NEO's five dimensions. Yet while, on the one hand, it's undoubtedly the case that emotional stability and coolness under pressure can sometimes tip the balance in professions where focus and levelheadedness have their say (the cockpit and the operating theater being just two cases in point), it should also be remembered that the marriage between Neuroticism and creativity is an enduring one. Some of art and literature's greatest offerings down the ages have been mined, not in the shallow waters of the brain's coastal perimeters, but in the deep, uncharted labyrinths of the soul.

But if occupational psychologists have uncovered individual differences in temperament based upon models of job performance—axes of personality that code for success in the workplace—how does the psychopath get along? In 2001, Donald Lynam and his colleagues at the University of Kentucky conducted a study to find out, and discovered that their unique personality structure conceals a telltale configuration of traits, as ruthless as it is mesmeric. Lynam asked a group of the world's top psychopathy experts (fellow academics with a proven track record in the field) to rate, on a scale of 1 to 5 (1 being extremely low, 5 extremely high) how they thought psychopaths measured up on a series of thirty sub-traits—the constituent parts of each of the primary dimensions that comprise the Big Five. The results are shown in figure 2.5.

OPENNESS TO EXPERIENCE	CONSCIENTIOUSNESS	EXTRAVERSION
Fantasy 3.1	Competence 4.2	Warmth 1.7
Aesthetics 2.3	Order 2.6	Gregariousness 3.7
Feelings 1.8	Dutifulness 1.2	Assertiveness 4.5
Actions 4.3	Achievement Striving 3.1	Activity 3.7
Ideas 3.5	Self-Discipline 1.9	Excitement Seeking 4.7
Values 2.9	Deliberation 1.6	Positive Emotions 2.5

AGREEABLENESS	NEUROTICISM
Trust 1.7	Anxiety 1.5
Straightforwardness 1.1	Angry Hostility 3.9
Altruism 1.3	Depression 1.4
Compliance 1.3	Self-Consciousness 1.1
Modesty 1.0	Impulsiveness 4.5
Tender-Mindedness 1.3	Vulnerability 1.5

Figure 2.5. Experts' ratings of the psychopathic personality profile as revealed by performance on the Big Five (Miller et al., 2001)

As we can see, the experts have the psychopaths just about flatlining when it comes to Agreeableness, which is not surprising given that lying, manipulation, callousness, and arrogance are pretty much considered the gold standard of psychopathic traits by most clinicians. Conscientiousness ratings are nothing to write home about either. Impulsivity, lack of long-term goals, and failure to take responsibility are up there, as we'd expect. But notice how Competence bucks the trend—a measure of the psychopath's unshakable self-confidence and insouciant disregard for adversity—and how the pattern continues with Neuroticism: Anxiety, Depression, Self-Consciousness and Vulnerability barely show up on the radar, which, when combined with strong outputs on Extraversion (Assertiveness and Excitement Seeking) and Openness to Experience (Actions), generates that air of raw, elemental charisma.

The picture that emerges is of a profoundly potent, yet darkly quicksilver personality. Dazzling and remorseless on the one hand. Glacial and unpredictable on the other.

The picture of a U.S. president? At first one might think, maybe not. But in 2010, Scott Lilienfeld teamed up with forensic psychologist Steven Rubenzer and Thomas Faschingbauer, professor of psychology at the Foundation for the Study of Personality in History, in Houston, Texas, and helped them analyze some rather interesting data. Back in 2000, Rubenzer and Faschingbauer had sent out the NEO Personality Inventory to the biographers of every U.S. president in history.* It included questions such as "You should take advantage of others before they do it to you." And "I never feel guilty over hurting people." In total, there were 240 of these items. Plus a catch. It wasn't the biographers who were being tested. But their subjects. The biographers, based on their knowledge, had to answer on their subjects' behalf.

* In actual fact, the NEO formed part of a larger, 592-item questionnaire that assessed a wide range of variables including personality, intelligence, and behavior. However, statistical techniques make it possible to extrapolate a psychopathic personality profile from an individual's overall performance on the NEO.

The results made interesting reading. A number of U.S. presidents exhibited distinct psychopathic traits, with John F. Kennedy and Bill Clinton leading the charge.* Not only that, but just look at how the Roosevelts fare. Some of history's golden boys are up there in the mix.

So should we be overly worried? Should it be cause for concern when the head of the most powerful nation on earth shares, as Jim Kouri noted, a significant proportion of his core personality traits with serial killers? Maybe. But to see where Lilienfeld, Rubenzer, and Faschingbauer are coming from with their political personality profiles, we need to dig deeper into precisely what it means to be a psychopath.

When Personality Goes Wrong

You need to be very careful when talking about personality disorder. Because everyone's got one, right? So let's get it straight from the start: personality disorders are not the preserve of those who piss you off (a common misconception among narcissists). Instead, as the *Diagnostic and Statistical Manual of Mental Disorders*† defines them, they are "an enduring pattern of inner experience and behavior that deviates markedly from the expectations of the culture of the individual who exhibits it."

The key word here is *enduring*. A personality disorder is not just for Christmas (though Christmas does, admittedly, bring out the best in them). No, personality disorders are characterized by deeply ingrained, inflexible patterns of thinking, feeling, or relating to others, or by the inability to control or regulate impulses that cause distress

* The table appears in www.wisdomofpsychopaths.com.

† The *Diagnostic and Statistical Manual of Mental Disorders* (*DSM*), published by the American Psychiatric Association, provides a common language and standard criteria for the classification of mental disorders. It is used in the United States, and in varying degrees around the world, by clinicians and researchers alike—as well as by pharmaceutical and health insurance companies, and by psychiatric drug regulation agencies. The manual was first published in 1952. The latest version, DSM-IV-TR, was published in 2000. DSM-V is due to be published in May 2013.

or impaired functioning. They may not be exclusive to those who piss you off. But if someone's got one, they will.

DSM classifies personality disorders into three distinct clusters.* There's odd/eccentric, dramatic/erratic, and anxious/inhibited. And, believe you me, they're all there. The cat-infested, crystal-gazing aunt with the tea-cozy hat and the big, dangly earrings, who thinks her bedroom is teeming with "presences" and that the pair across the road are aliens (schizotypal); the bling-toting, permatanned pool attendant, who's had so much Botox he makes even Mickey Rourke look normal (narcissistic); and the cleaning lady I once hired, who, after three excruciating hours, was still working on the damn bath, for Christ's sake (obsessive-compulsive). (I was paying her by the hour. So who was the crazy one there? I wonder.)

But personality disorders don't just cause trouble in everyday life. They draw a good deal of fire within clinical psychology, too. One bone of contention revolves around the word "disorder." With an estimated 14 percent of the general population diagnosed with one, the question arises as to whether, in fact, we should be calling them "disorders" at all. Might not, in reality, "personalities" be a better description? Well, maybe. But perhaps we should ask what personality disorders are, exactly. Do they, for instance, comprise a separate archipelago of pathology, epidemiologically adrift off the coast of mainland personality? Or do they, in contrast, form part of the Big Five peninsula: remote outposts of temperament at its darkest, most storm-battered fringes?

Support for this latter, anti-separationist view comes from a wide-ranging survey conducted by Lisa Saulsman and Andrew Page in 2004. Saulsman and Page scoured the clinical literature—studies that looked, in turn, at the relationship between each of the ten personality disorders listed in DSM on the one hand, and each of the Big Five personality dimensions on the other—and chucked what they found into one big melting pot of data. Analysis revealed that all ten personality disorders can be accounted for within the framework of

* For the complete list of disorders see www.wisdomofpsychopaths.com.

the Big Five model. But, crucially, it was an overriding "Big Two" that did most of the heavy lifting: Neuroticism and Agreeableness.

To illustrate, Saulsman and Page found that disorders particularly characterized by emotional distress (e.g., Paranoid, Schizotypal, Borderline, Avoidant, and Dependent) display strong associations with Neuroticism, while those typified by interpersonal difficulties (e.g., Paranoid, Schizotypal, Antisocial, Borderline, and Narcissistic) fall down, perhaps not surprisingly, on Agreeableness. Also implicated, but to a somewhat lesser degree, were the dimensions of Extraversion and Conscientiousness. Disorders either side of what we might call the Socialite-Hermit divide (Histrionic and Narcissistic, on one; Schizoid, Schizotypal, and Avoidant on the other) posted, respectively, high and low scorecards on Extraversion. Those either side of the Easy Rider–Control Freak border (Antisocial and Borderline in one camp versus Obsessive-Compulsive in the other) were similarly bipolar when it came to Conscientiousness.

The case seems pretty convincing. If the omnipotent Big Five comprise our personality solar system, then the rogue constellation of disorders certainly forms part of the firmament. But where, once again, does that leave psychopaths?

The Mask of Sanity

Psychopathy—like personality itself—first appears on the radar, in exquisitely mischievous though wholly unmistakable form, amidst the musings of the ancient Greeks. The philosopher Theophrastus (c. 371–287 B.C.), the successor to Aristotle as head of the Peripatetic school in Athens, delineates, in his book *The Characters*, a coruscating caseload of thirty moral temperaments. One of the assembled rings several cacophonous bells.

"The Unscrupulous Man," Theophrastus laments, "will go and borrow more money from a creditor he has never paid . . . When marketing he reminds the butcher of some service he has rendered him, and, standing near the scales, throws in some meat, if he can, and a soupbone. If he succeeds, so much the better; if not, he will snatch a piece of tripe and go off laughing." And go off laughing he did. But fast-forward a couple of thousand years, to the early nineteenth century, and the unscrupulous man returns, this time as one of the key metaphysical players in the debate over free will. Could it possibly be the case, philosophers and physicians conjectured, that certain moral transgressors, certain unconscionable ne'er-do-wells, weren't simply "bad," but were, in fact, in contrast to other miscreants, possessed of little or no understanding of the consequences of their actions? One of them certainly thought so.

In 1801, a French physician by the name of Philippe Pinel scribbled in his notebook the words *manie sans délire* after looking on in horror as a man coolly, calmly, and collectedly kicked a dog to death in front of him. Later in that same year, Pinel was to compile a meticulous, comprehensive—and, to this day, highly accurate—account of the syndrome. Not only had the man in question exhibited not the slightest flicker of remorse for his actions, he had also, in most other respects, appeared perfectly sane. He seemed, to coin a phrase that many who have since come into contact with psychopaths concur with, to be "mad without being mad." *Manie sans délire.*

The Frenchman, it turned out, wasn't alone in his ponderings. The physician Benjamin Rush, practicing in America in the early 1800s, provides accounts similar to Pinel's, of equally abhorrent behaviors and equally untroubled thought processes. To the perpetrators of such actions, Rush accords an "innate preternatural moral depravity," in which "there is probably an original defective organization in those parts of the body, which are occupied by the moral faculties of the mind."

The will, he continues, might be deranged even in "many instances of persons of sound understandings . . . the will becom[ing] the involuntary vehicle of vicious actions, through the instrumentality of the passions."

He anticipated modern neuroscience by a couple of hundred years. The neural tsunami of madness need not, in other words, wash apocalyptically up on the crystalline shores of logic. You can be sound of mind and "unsound," simultaneously.

Spool forward a century and a half, and across the Atlantic, at the Medical College of Georgia, the American physician Hervey Cleckley provides a more detailed inventory of *la folie raisonnante*. In his book *The Mask of Sanity*, published in 1941, Cleckley assembles the following somewhat eclectic identikit of the psychopath. The psychopath, he observes, is an intelligent person, characterized by a poverty of emotions, the absence of a sense of shame, egocentricity, superficial charm, lack of guilt, lack of anxiety, immunity to punishment, unpredictability, irresponsibility, manipulativeness, and a transient interpersonal lifestyle—pretty much the picture that twenty-first-century clinicians have of the disorder today (though with the aid of lab-based research programs, and the development of techniques such as EEG and fMRI, we're now beginning to get a better understanding as to why). But interspersed in Cleckley's portrait are the brushstrokes of what looks like genius. The psychopath is described as having "shrewdness and agility of . . . mind," as "talk[ing] entertainingly" and possessing "extraordinary charm."

In a memorable passage, Cleckley describes the innermost workings of the minds of these social chameleons, the day-to-day life behind the icy curtain of unfeeling:

> [The psychopath] is unfamiliar with the primary facts or data of what might be called personal values and is altogether incapable of understanding such matters. It is impossible for him to take even a slight interest in the tragedy or joy or the striving of humanity as presented in serious literature or art. He is also indifferent to all these matters in life itself. Beauty and ugliness, except in a very superficial sense, goodness, evil, love, horror, and humor have no actual meaning, no power to move him . . . He is, furthermore, lacking in the ability to see that others are moved. It is as though he were colorblind, despite his sharp intelligence, to this aspect of human existence. It cannot be explained to him because there is nothing in his orbit of awareness that can bridge the gap with comparison. He

> can repeat the words and say glibly that he understands, and there is no way for him to realize that he does not understand.

The psychopath, it's been said, gets the words, but not the music, of emotion.

I got a distinct taste of what Cleckley was driving at in one of my very first encounters with a psychopath. Joe was twenty-eight, better-looking than Brad Pitt, and had an IQ of 160. Why he'd felt the need to beat that girl senseless in the parking lot, drive her to the darkness on the edge of that northern town, rape her repeatedly at knifepoint, and then slit her throat and toss her facedown in that Dumpster in a deserted industrial park is beyond comprehension. Parts of her anatomy were later found in his glove compartment.

In a soulless, airless interview suite smelling faintly of antiseptic, I sat across a table from Joe—a million miles, and five years, on from his municipal, blue-collar killing field. I was interested in the way he made decisions, the stochastic settings on his brain's moral compass—and I had a secret weapon, a fiendish psychological trick up my sleeve, to find out. I posed him the following dilemma:

A brilliant transplant surgeon has five patients. Each of the patients is in need of a different organ, and each of them will die without that organ. Unfortunately, there are no organs currently available to perform any of the transplants. A healthy young traveler, just passing through, comes into the doctor's office for a routine checkup. While performing the checkup, the doctor discovers that the young man's organs are compatible with all five of his dying patients. Suppose, further, that were the young man to disappear, no one would suspect the doctor. Would the doctor be right to kill the young man to save his five patients?

This moral conundrum was first put forward by Judith Jarvis Thomson, the author of the fat-man-and-trolley experiment we discussed in chapter 1. Though certainly a talking point, it's pretty easily resolved by most people. It would be morally reprehensible for the doctor to take the young man's life—no physician has the right to kill a patient, irrespective of how humane or compassionate the justification may seem at the time. It would be murder, plain and simple. But what would someone like Joe's take on it be?

"I can see where the problem lies," he commented matter-of-factly when I put it to him. "If all you're doing is simply playing the numbers game, it's a fucking no-brainer, isn't it? You kill the guy, and save the other five. It's utilitarianism on crack . . . The trick's not to think about it too much . . . If I was the doctor, I wouldn't give it a second thought. It's five for the price of one, isn't it? Five bits of good news—I mean, what about the families of these guys?—against one piece of bad. That's got to be a bargain. Hasn't it?"

"They do emotions by numbers," one senior forensic psychiatrist told me as we sat in his office talking about psychopaths. In Joe's case, it would seem, quite literally.

Identity Crisis

The psychopath's powers of persuasion are incomparable; their psychological safecracking abilities, legendary. And Joe, the killer, the rapist, with his arctic blue stare and genius-level IQ, was certainly no exception to the rule. Sometimes, in fact, when you talk to a psychopath in interview, it can be difficult to believe that there's anything wrong at all—if you don't know any better. Which is just one of the reasons why coming up with a precise classification of the disorder on which everyone is agreed has proven so tricky down the years.

It's been three decades now since psychopathy got its clinical residence permit. In 1980, Robert Hare (whom we met in chapter 1) unveiled the Psychopathy Checklist, the inaugural (and to many, still the best) test for detecting the presence of the disorder. The checklist—which, in 1991, underwent a facelift: it's since been renamed the Psychopathy Checklist–Revised (PCL-R)—comprises a twenty-item questionnaire carrying a maximum score of 40 (on each item, an individual can score either 0, "doesn't apply"; 1, "applies somewhat"; or 2, "fully applies"), and was developed by Hare on the basis of both his own clinical observations, and those previously identified by Hervey Cleckley in Georgia.

Most of us score around 2. The entry level for psychopaths is 27.*

Perhaps not surprisingly, given the way personality theorists like to do things, the 20 items that make up the PCL-R have, on numerous occasions, just like the 240 items that comprise the NEO, been subjected to the statistical card-shuffling game that is factor analysis. The results of the game have varied over the years, but recent activity by a number of clinical psychologists suggests that, in exactly the same way that there exist five main dimensions of personality space in general, there lurk four main dimensions in the spectral psychopath nebula nested mercurially within it (see figure 2.6).

INTERPERSONAL ITEMS	AFFECTIVE ITEMS	LIFESTYLE ITEMS	ANTISOCIAL ITEMS
Glibness/superficial charm	Lack of remorse or guilt	Need for stimulation / proneness to boredom	Poor behavioral controls
Grandiose sense of self-worth	Shallow affect	Parasitic lifestyle	Early behavioral problems
Pathological lying	Callous/lacks empathy	Lack of realistic, long-term goals	Juvenile delinquency
Conning/manipulative	Failure to accept responsibility for own actions	Impulsivity	Revocation of conditional release
		Irresponsibility	Criminal versatility

Figure 2.6. A four-factor model of the PCL-R (from Hare, 2003)

* The PCL-R is issued in clinical settings, by qualified personnel, and is scored on the basis of an extensive file review and a semi-structured interview. Do *not* try to use it on your bank manager.

Psychopathy, in other words, is a composite disorder consisting of multiple interrelated components that range discretely and independently along a number of different spectra: interpersonal, emotional, lifestyle, and antisocial—a witches' brew of personality leftovers.

But which of these spectra are most important? Is someone who scores high on the antisocial elements of the checklist, for example, and lower, say, on the interpersonal dimension, more or less of a psychopath than someone whose profile is the complete opposite?

Questions like these surface quite regularly in the battle for the psychopath psyche, in the empirical and diagnostic combat zones of clinical definition. Take, for instance, *DSM*'s listing of Antisocial Personality Disorder (ASPD), an area of particular strategic importance in the epidemiological conflicts. The official line, as set out by the American Psychiatric Association, is that ASPD and psychopathy are, in fact, synonymous. ASPD is defined as "a pervasive pattern of disregard for, and violation of, the rights of others that begins in childhood or early adolescence and continues into adulthood." The individual must be age eighteen or over, show evidence of conduct disorder* before the age of fifteen, and present with at least three of the following criteria:

1. Failure to conform to social norms with respect to lawful behaviors, as indicated by repeatedly performing acts that are grounds for arrest
2. Deceitfulness, as indicated by repeated lying, use of aliases, or conning others for personal profit or pleasure

* Conduct disorder (CD), according to *DSM*, is characterized by "a repetitive and persistent pattern of behavior in which the basic rights of others or major age-appropriate societal norms or rules are violated . . . manifested by the presence of three (or more) of the following criteria in the past 12 months, with at least one criterion present in the past 6 months: aggression to people and animals . . . destruction of property . . . deceitfulness or theft . . . serious violation of rules." In addition, CD should result in "clinically significant impairment in social, academic, or occupational functioning." Two forms of CD are specified: childhood-onset (in which at least one criterion of the disorder must be in evidence prior to the age of ten); and adolescent-onset (in which no criteria should have occurred prior to the age of ten).

3. Impulsivity, or failure to plan ahead
4. Irritability and aggressiveness, as indicated by repeated physical fights or assaults
5. Reckless disregard for safety of self or others
6. Consistent irresponsibility, as indicated by repeated failure to sustain consistent work behavior or honor financial obligations
7. Lack of remorse, as indicated by being indifferent to, or rationalizing, having hurt, mistreated, or stolen from another.

But is this really the same thing as psychopathy? Many theorists argue not—and that although there is certainly overlap between the two, the fundamental difference lies in insidious vagaries of emphasis: in the manifest imbalance between the welter of behavioral items of "socially deviant" criteria that characterize ASPD and the core affective impairment, the shadowy emotional twilight, redolent of the psychopath.

The ramifications, statistical or otherwise, are not without consequence. In prison populations ASPD is the psychiatric equivalent of the common cold, with as many as 80 to 85 percent of incarcerated criminals, according to Robert Hare, meeting the requirements for the disorder. Contrast this with just a 20 percent hit rate for psychopaths. In addition, this 20 percent minority punches well above its weight. Around 50 percent of the most serious crimes on record—crimes such as murder and serial rape, for instance—are committed by psychopaths, and continue to be committed by psychopaths.

Studies comparing the recidivism rates among psychopathic and non-psychopathic prisoners reveal that the former are up to three times more likely to reoffend than the latter within a period of just one year. If we factor violence into the equation, the curve gets even steeper. The psychopath emerges as anything up to five times more likely to beat, rape, kill, or mutilate his way back behind bars. More accurate is to say that the relationship between ASPD and psychopathy is asymmetrical. For every four people diagnosed with ASPD, you may also have a psychopath on your hands. But every individual presenting with psychopathy will also, by default, be presenting with ASPD.

Killer Difference

To demonstrate a little more clearly, perhaps, the difference between the two syndromes, consider the following two case histories:

CASE 1

Jimmy is thirty-four years old and has been sentenced to life imprisonment for murder. He's always had a short fuse, and got involved in a fight in a pub that ended in a fatal head injury. Generally speaking, Jimmy is popular in prison, keeps his nose clean and his head down. First impressions of him are of an immature, happy-go-lucky kind of guy who gets on well with staff and fellow prisoners alike.

Jimmy's criminal record (which consists of around half a dozen offenses) began at the age of seventeen when he was arrested for shoplifting—though before that, according to his parents, things were already going downhill. A couple of years earlier, when he was fifteen, Jimmy began drifting into trouble both at home and at school. He started staying out late at night, joined a notorious local gang, lied habitually, got into fights, stole cars, and vandalized property. When he turned sixteen, Jimmy dropped out of school and began working for a well-known department store, loading trucks. He also began to drink heavily, occasionally stealing from the warehouse to "tide him over." He had trouble holding on to his money, and making ends meet frequently posed a challenge, so he started dealing marijuana. A couple of years later, three months after his eighteenth birthday, he wound up on probation and moved in with his girlfriend.

After losing his job, and subsequently several others, Jimmy found work at a garage. Despite constant arguments over his drinking, drug dealing, and spending habits, his relationship with his girlfriend remained pretty much on track for a while. There were a couple of affairs, but Jimmy put an end to both of them. He felt guilty, he said. And he was also concerned that his girlfriend would find out, and leave him.

Then the drinking started getting out of hand. One night, in his local pub, Jimmy got in a fight. The bar staff intervened quickly and Jimmy was shown the door. Normally, he would've gone quietly. But this time, for some reason, he just couldn't "let it go." So he picked up a pool cue and smashed it—with so much force it shattered—over the other guy's head from behind: a blow, unfortunately, which caused a massive brain hemorrhage. The police arrived. And Jimmy confessed on the spot. At his trial, he pleaded guilty.

CASE 2

Ian is thirty-eight and is serving a life sentence for murder. One night he pulled into a motel to get something to eat and ended up shooting the receptionist at point-blank range in order to steal the money from her till. In prison, he's known to be heavily involved in both taking and dealing drugs—plus quite a few other forms of racketeering. He is charming and upbeat to talk to—at least, that is, to start with. But conversations usually end up taking a violent or sexual turn, a fact not lost on female members of the staff. He's had a number of jobs on the wing since being admitted, but his unreliability, combined with his explosive aggression (often when he fails to get his own way), have led to a checkered employment history. Ask fellow prisoners what they think of him and most of them admit to a mixture of fear and respect. It's a reputation he enjoys.

Ian's criminal record begins at the age of nine, when he stole some computer equipment from his local youth club. It quickly escalated into the attempted murder of a classmate when he was eleven. When confronted by Ian in the school lavatories, the boy refused to hand over his dinner money—so Ian put a plastic bag over his head and attempted to suffocate him in one of the cubicles. But for the intervention of a teacher, Ian says, he would have "made sure the fat bastard never needed his dinner money again." Recalling the incident, he shakes his head and smiles.

On leaving school, Ian spent most of his time checking in and out of various secure units. His criminal proclivities were versatile, to say the least: deception; shoplifting; burglary; street robbery; grievous bodily harm; arson; drug dealing; pimping. Unable to hold down a job for more than a couple of weeks at a time, he either sponged off friends or lived off the proceeds of his crimes. He enjoyed a transient existence, drifting from sofa to sofa and hostel to hostel, preferring to move around instead of putting down roots. Since he exuded a confident, charming, and self-assured persona, there was always someone willing to put a roof over his head—usually "some woman" he'd chatted up in a bar. But inevitably it ended in tears.

Ian has never been married, but has had a string of live-in girlfriends. His longest relationship lasted six months, and like all the others, it was peppered with violent rows. On each occasion, it was Ian who moved into his partner's place, rather than the other way round. And, on each occasion, he "swept them off their feet." Affairs were commonplace. In fact, Ian has trouble remembering a time when he didn't, as he puts it, "have more than one chick on the go"—though he claims he was never unfaithful. "Most of the time I came back to her at night," he says. "What more do they want?"

At his trial, the evidence against Ian was ironclad. Yet he entered a plea of not guilty, and still, to this day, maintains his innocence. As the verdict was read out in court, he smiled in the direction of the victim's family and gave the judge the finger as he was escorted from the dock. Since being in prison, Ian has filed two appeals against his sentence. He is supremely confident, despite repeated protestations to the contrary from his solicitor, that his case will be reviewed and that the verdict will be overturned. The champagne's on ice, he says.

So you're the clinician, and Ian and Jimmy are cellmates. They're sitting in the corridor awaiting consultation. Do you think you could identify the psychopath out of the two? On the surface, it might be difficult. But let's look again at the criteria for ASPD. Both show a failure to conform to social norms. And both have tendencies toward poor behavioral control—toward impulsivity, aggressiveness, and irresponsibility. A clear-cut diagnosis, I'd suggest.

But now let's examine the psychopathic narrative. The need for stimulation and a parasitic lifestyle? More Ian's bag than Jimmy's, I would say. Yet it's when we come to emotion, or more specifically the lack of it, that Ian's "mask of sanity" really begins to slip. Charming, grandiose, manipulative, deficient in empathy and guilt: Ian is so good at psychopathy, it's almost as if he's been practicing. As if he's recently come out of some secret psychopath finishing school—with honors. ASPD is psychopathy with added emotion. Psychopathy is an emotionless void.

Criminal Omission

That psychopathy doesn't pass muster with the custodians of *DSM* is an intriguing act of omission. The reason most cited for its curious and conspicuous exclusion is one of empirical intractability—that, and its supposed synonymy with ASPD. Guilt, remorse, and empathy are not, perhaps, the most quantifiable of constructs to grapple with. So best, instead, to stick to observable behavior, lest the specter of subjectivity rear its head.

This is problematic, to say the least. For a start, studies reveal that concordance rates among clinicians are actually pretty high when it comes to the PCL-R. The scale, to use the proper terminology, has good "inter-rater reliability." And besides, as one senior psychiatrist told me, "you can smell a psychopath within seconds of them coming through the door."

But that's not the only bone of contention. The enigma of the psychopath's identity, of what, precisely, the mask of sanity conceals, is given another phenomenological twist by an unnerving observation a little closer to home. Not all psychopaths are behind bars. The majority, it emerges, are out there in the workplace. And some of them, in fact, are doing rather well. These so-called successful psychopaths—like the ones Scott Lilienfeld studies—pose a problem for ASPD and, incidentally, for proponents of the PCL-R.

A recent study led by Stephanie Mullins-Sweatt at Oklahoma State University presented attorneys and clinical psychologists with a prototypical description of a psychopath. After reading the profile, the two groups of professionals were put on the spot. Were they, Mullins-Sweatt wanted to know, able to call to mind anyone they knew, past or present, who, in their own personal opinion, was befitting of such a description (and who, needless to say, was successful in their given career)? If so, could they rate that person's personality on a test of the Big Five?

The results made interesting reading. Consistent with expectation, the successful psychopaths—conjured, among others, from the worlds of business, academia, and law enforcement*—emerged as nefarious and dastardly as ever. Just like their unsuccessful counterparts, they were described in general terms as being "dishonest, exploitative, low in remorse, minimizing of self-blame, arrogant, and shallow."

No surprises there. But when it came to the Big Five, the similarity continued. Just as in Donald Lynam's study, where the experts donned their rating caps, the successful psychopaths, like their prototypical alter egos, are portrayed (hypothetically) as being high on dimensions of assertiveness, excitement-seeking, and activity . . . and low on dimensions of agreeableness, such as altruism, compliance, and modesty. Moreover, with the exception of self-discipline (on which the unsuccessful psychopaths bombed and the successful ones excelled), conscientiousness profiles are also seen to converge, with both groups maxing on competence, order, and achievement-striving.

* "Top-notch police detective"; "dean from a major university"; "successful retail business"; "made large sum of money and was mayor for three years"; "managerial position in a government organization"; "endowed professor with numerous federal grants"—these are just some of the success indicators that surfaced in the study.

All of which begs the question: Where does the crucial difference lie? Does the fulcrum of disparity between successful and unsuccessful psychopaths, between presidents and pedophiles, pivot solely around self-discipline? Everything else being equal, such a possibility might actually hold some water. The ability to delay gratification, to put on hold the desire to cut and run (and also, needless to say, to run and cut), might well tip the balance away from criminal activity toward a more structured, less impulsive, less antisocial lifestyle.

Except that the question of criminal activity raises issues of its own. In both the revised psychopathy checklist—the PCL-R—and the criteria for antisocial personality disorder set out in DSM, "criminal versatility" and "repeatedly performing acts that are grounds for arrest" constitute, respectively, core diagnostic determinants of psychopathy. Symptoms, in other words. And yet, as the study by Mullins-Sweatt illustrates, neither of these items need necessarily apply to the successful branch of the species. It's perfectly possible to be a psychopath and not a criminal.

So do successful psychopaths fall short of the real deal? Are they a neuron short of a synapse compared to their more notorious, nefarious namesakes? It's a tricky one to call. But fifteen years ago, in an attempt to do just that, one man stepped up to the plate—and wound up with me and a mountain of alligator tacos in a diner in downtown Atlanta.

The Low Road Through Psychopathy

In 1996, Scott Lilienfeld and his collaborator Brian Andrews were in the process of grappling with exactly this conundrum. As an experienced researcher in the field, with quite a few psychopaths already under his belt, Lilienfeld had come to a definitive, if perplexing, conclusion. Insofar as the inaugural constitution of the disorder was concerned—the traditional concept of what it really meant to be a psychopath, as set out by founding father Hervey Cleckley—the PCL-R, and other clinical measures, were themselves behaving rather oddly. Over the years, Lilienfeld realized, the diagnostic spotlight had widened. Initially focused on the personality traits that underpinned the disorder, the emphasis now seemed to be as much, if not more, on antisocial acts. The psychopath circus had gotten stuck in the mud of forensics.

As a case in point, Lilienfeld and Andrews cited fearlessness. In his original manifesto back in 1941, Cleckley had contended that low anxiety levels constituted one of the psychopath's true calling cards: a cardinal feature of the syndrome. Yet where, precisely, did that show up in the fabric of the PCL-R? Behind such omissions, Lilienfeld detected a major theoretical fault line developing between the ways different sectors of the clinical and research fraternities were coming to view psychopathy: an old-school divide between two analytical traditions—between qualitative psychological means and quantitative behavioral ends.

Two camps, it appeared, had crept out of the epistemological woodwork. In one were the Cleckleyites, whose main area of interest lay in the undercoat of personality; in the other were ensconced the behaviorists, beholden to DSM and the gospel of ASPD, who tended, in contrast, to focus on the criminal record. Such a schism, needless to say, was conducive to neither coherent empirical inquiry nor diagnostic consensus. An individual who, on the one hand, possessed all the necessary requirements of the psychopathic personality, but who didn't, on the other, partake in recurrent antisocial behavior—one of Mullins-Sweatt's "subclinical" variety, for example—would be endorsed as a psychopath by advocates of the personality-based approach, but would, Lilienfeld and Andrews discerned, be turned smartly away at the door by their behaviorist, actions-speak-louder-than-words opposite numbers.

And the dynamic cut both ways. As we saw with Ian and Jimmy, not everyone who engages in habitual criminal activity is a psychopath. It's just a small minority, in fact. Something needed to be done to assimilate the rival frameworks, to bring these yawningly different perspectives into alignment. And Lilienfeld and Andrews had the answer.

The Psychopathic Personality Inventory (PPI for short) consists of 187 questions. It's not exactly the snappiest questionnaire in the world. But then again, the nature of its subject matter isn't snappy, either. Eight separate personality dimensions converge in this psychometric behemoth, making it one of the most comprehensive tests of psychopathy yet devised. Interestingly, our old friend factor analysis uncovers a familiar pattern. These eight independent satellite states of the psychopathic personality—Machiavellian Egocentricity (ME); Impulsive Nonconformity (IN); Blame Externalization (BE); Carefree Nonplanfulness (CN); Fearlessness (F); Social Potency (SOP); Stress Immunity (STI); and Coldheartedness (C)—divide and re-form along three superordinate axes . . .

1. Self-Centered Impulsivity (ME + IN + BE + CN)
2. Fearless Dominance (SOP + F + STI)
3. Coldheartedness (C)

. . . to reveal, in the statistical residue, once the mathematical dust clouds have settled, the structural DNA of pure, unadulterated psychopathy. This was the genome that Cleckley had originally sequenced, untarnished by time, unsullied by counts of transgression. And anyone, pretty much, could prove a positive match.

The tequila is flowing. And so, as we scarf down the tacos, is Lilienfeld, as he explains what it really means, in terms of the nuclei of core personality, to be deemed to be a psychopath. He recounts the empirical rationale behind the development of the PPI: "The problem at the time with the existing measures of the syndrome was that most of them had been honed on criminal or delinquent populations. Yet we know that people with psychopathic traits function perfectly well on the 'outside'—and that some of them are extremely successful. Ruthlessness, mental toughness, charisma, focus, persuasiveness, and coolness under pressure are qualities, so to speak, that separate the men from the boys, pretty much across the board. So somehow we had to bridge the gap between the incarcerated 'forensic' psychopaths and their elite, high-functioning counterparts. The high road through psychopathy was well established. But what about a low road . . . ?

"We reasoned that psychopathy was on a spectrum. And it goes without saying that some of us will be high on some traits, but not on others. You and I could post the same overall score on the PPI. Yet our profiles with regard to the eight constituent dimensions could be completely different. You might be high on Carefree Nonplanfulness but correspondingly low on Coldheartedness, whereas for me it might be the opposite."

Lilienfeld's notion of psychopathy being on a spectrum makes a good deal of sense. If psychopathy is conceptualized as an extension of normal personality, then it follows logically that psychopathy itself must be scalar, and that more or less of it in any given context might confer considerable advantages. Such a premise is not without precedent in the annals of mental dysfunction (if, indeed, psychopathy *is* dysfunctional, given its benefits under certain conditions). The autistic spectrum, for instance, refers to a continuum of abnormality in social interaction and communication ranging from severe impairments at the "deep end"—those who are silent, mentally disabled, and locked into stereotypical behaviors such as head rolling or body rocking, for example—to mild interference at the "shallow end": high-functioning individuals with active, but distinctly odd, interpersonal strategies, narrowly focused interests, and an undue preoccupation with "sameness," rules, and ritual.

Less familiar, perhaps, but equally pertinent, is the schizophrenic spectrum. Research on the construct of schizotypy suggests that psychotic experiences of one form or another (usually of the harmless and nondistressing variety) are relatively common among the general population, and that rather than being seen as a unitary condition—you've either got it or you haven't—schizophrenia should be viewed as a dimensional disorder, with arbitrary cutoffs between normal, odd, and ill. Within this framework, the symptoms of Schizotypal Personality Disorder (odd beliefs; bizarre speech patterns; eccentric interpersonal style) are very much construed as the nursery slopes of the central schizophrenia massif. Exactly as with psychopathy, at low to medium altitudes the "disorder" is perfectly manageable—beneficial, even, in some contexts (the link between schizotypy and creativity is well established). But above and beyond the snow line, conditions get ever more hazardous.

Such an approach to the conundrum of mental disorder has an intuitive, commonsense appeal. That nagging supposition that we're all just a little bit bonkers is a difficult one to ignore. Yet when it comes to psychopathy and the dimensional denouement of a psychopathic spectrum, Scott Lilienfeld certainly hasn't had things entirely his own way. There are those who take issue with his sliding-scale solution and have evidence of their own to throw at it. Foremost among them is a man named Joseph Newman.

What You Don't Know Can't Hurt You

Joe Newman is professor of psychology at the University of Wisconsin–Madison, and an hour in his office is like sitting in a psychological wind tunnel—like a whitewater raft ride through the rapids of cognitive science. For the better part of thirty years now, Newman has been in and out of some of the toughest prisons in the Midwest. Not, of course, as an inmate, but as one of the world's most intrepid researchers, working with psychopaths high above the snow line of dysfunction. Though long since acclimated to the harsh, unforgiving conditions, he concedes there are times, even now, when it all gets a little bit hairy.

He recalls, for example, an incident that occurred a few years back with a guy who scored 40 on the PCL-R. That, if you recall, is the maximum you can get. And it's rare. The guy was a "pure" psychopath. "Usually there's a point in the interview where we like to push people a little bit," Newman tells me. "You know, challenge them. Gauge their reaction. But when we did that with this fella—and he was a really nice guy up until then: charming, funny, big personality—he got this kind of cold, derelict look in his eyes, difficult to describe, but you know it when you see it, which just seemed to say, 'Back off!' And you know what? We did! He scared the absolute shit out of us."

Newman, by his own admission, sometimes gets that look in his own eyes. He stops short of saying that it takes one to know one. But growing up as a kid on the mean streets of New York, he's had knives, and guns, and the whole nine yards pulled on him. Without a trace of irony, he's grateful, he says. It was a taste of things to come. In academia.

Newman is more abstemious than most when it comes to the selection criteria for a psychopath. "My main concern is that the label [of psychopath] is applied too liberally, and without sufficient understanding of the key elements," he purrs in a soft, almost apologetic tone. "As a result, the doors are thrown open to pretty much anyone, and the term is often applied to ordinary criminals and sex offenders whose behavior may reflect primarily social factors or other emotional problems that are more responsive to treatment than psychopathy."

Similarly, he's more than amenable to the idea of psychopaths coexisting outside the criminal firmament, often doing very well for themselves in professions that might otherwise come as a surprise to those less well versed in the building blocks of the psychopathic personality: as surgeons, lawyers, and corporate head honchos, for example. "The combination of low risk aversion and lack of guilt or remorse, the two central pillars of psychopathy," he elucidates, "may lead, depending on circumstances, to a successful career in either crime or business. Sometimes both."

So no problems there. But where Newman does go against the grain is when it comes to the underlying cause, or etiology, of the disorder. Conventional theoretical wisdom holds that psychopaths are incapable of experiencing fear, empathy, and a host of other emotions, which anesthetizes their social cognition, and which in turn renders them commensurably incapable of countenancing such feelings in those they come into contact with. This position, taken by, among others, fellow psychopath czar James Blair at the National

Institute of Mental Health in Bethesda, Maryland, implicates neural dysfunction, specifically in relation to the amygdala, the brain's CEO of emotion, plus a number of structures closely connected with it—the hippocampus, superior temporal sulcus, fusiform cortex, anterior cingulate, and orbitofrontal cortex, for instance—as the primary cause of the syndrome, as the core biological basis behind the industry standard psychopathic dyad: the behavioral accompaniments of profound emotional impairment and repeated antisocial action.

But Newman has other ideas. Far from believing that psychopaths are incapable of fear—that they're the emotionless void that the literature traditionally paints them as—he proposes instead that actually they just don't notice it. Imagine, for instance, that you're an arachnophobe and the mere thought of anything with eight legs sends you into a cold sweat. Such being the case, a tarantula might, for all you know, be dangling a few centimeters above your head right at this very moment. But if you don't know it's there, you won't be afraid of it, will you? In your brain, it just doesn't exist.

In an ingenious experiment, Newman demonstrated that this might well be the story with psychopaths. Not just with spiders, but with most things. They don't feel distress or notice such emotion in others, because when they focus on a task that promises immediate reward, they screen everything "irrelevant" out. They get emotional "tunnel vision."

He and his coworkers presented a group of psychopaths and nonpsychopaths with a series of mislabeled images, such as those shown in figure 2.7.

Figure 2.7. The picture-word Stroop task (adapted from Rosinski, Golinkoff, and Kukish, 1975)

Their task, a favorite among cognitive psychologists, especially those interested in the mechanisms underlying attention, seems simple enough: name the picture while ignoring the incongruent word—against the clock, over a series of consecutive trials.

Most people, in fact, find this a tad tricky. The explicit instruction to name the focal image conflicts with the urge to read the discrepant word, a grinding of the gears that leads to hesitation. This hesitation, or "Stroop interference," as it's known (after J. R. Stroop, the man who came up with the original paradigm in 1935), is a measure of attentional focus. The faster you are, the narrower your attentional spotlight. The slower you are, the wider the arc of the beam.

If Newman's theory was to cut any ice and psychopaths really did suffer from the kind of information-processing deficit (or talent) that he was talking about, then it didn't take a rocket scientist to work out what should happen. They should be faster at naming the pictures than the non-psychopaths. They should zone in exclusively on the particular task at hand.

The results of the study couldn't have turned out better. Time and again, Newman found that while the non-psychopathic volunteers were completely undone by the discrepant picture-word pairings—taking longer to name the images—the psychopaths, in contrast, sailed through the task, pretty much oblivious to the jarring inconsistencies. What's more—and this is where things start to get a little sticky for Scott Lilienfeld and the psychopathic spectrum—Newman has detected an anomaly in the data: an abrupt discontinuity in response patterns once a critical threshold is reached. Everyone performs about the same, encounters the same degree of difficulty with such tasks, on the lower slopes of the PCL-R. But as soon as you hit psychopathy's clinical base camp, a score of 28 to 30, the dynamic dramatically changes. Indigenous populations at these rarer, higher altitudes suddenly find it easy. They just don't seem to process the glaring peripheral cues that, to everyone else, appear obvious.

And it's not that they're immune to them. Far from it. In a separate study, Newman and his colleagues presented psychopaths and non-psychopaths with a series of letter strings on a computer screen. Some of them were red. And some of them were green. And some of them were downright painful: volunteers were told that, following the random display of an arbitrary number of reds, they'd receive an electric shock. As expected, when their attention was cued away from the prospect of shock (i.e. when they were asked to state whether the letters appeared as upper- or lowercase), the psychopaths showed considerably less anxiety than the non-psychopaths. But incredibly, when the prospect of shock was made salient (i.e. when volunteers were explicitly asked to state what colors the letters appeared in, red or green), the trend, as Newman and his coauthors predicted, reversed. This time it was actually the psychopaths who got more edgy.

"People think [psychopaths] are just callous and without fear," he says. "But there is definitely something more going on. When emotions are their primary focus, we've seen that psychopathic individuals show a normal [emotional] response. But when focused on something else, they become insensitive to emotions entirely."

With a disconnect in response sets showing up at precisely the point on the PCL-R that things start to get clinical, the mystery as to what, precisely, psychopathy really is—whether it lies on a continuum or is a completely separate disorder—suddenly deepens.

Is psychopathy just a matter of degree? Or are the big boys in a league of their own?

One Small Step, One Giant Leap

It's reasonable to assume that the answer to such a question should, by its very nature, be black-and-white. That is, if psychopathy is on a continuum, then the trajectory from low to high, from Mother Teresa to John Wayne Gacy, must be linear, and the road to moral weightlessness smooth. And if not, not: you get the kind of precipitous increments in data patterns observed by Joe Newman. But actually, as anyone who has ever played the lottery will tell you, it's not that simple. The six winning numbers are certainly on a continuum: a continuum of 1 through 6. But the size of your winnings, from $1,000 to a $1,000,000 jackpot, is a different story entirely. The function is exponential, and the relationship between the numbers on a continuum on the one hand, and how they convert (quite literally in this case) to "real life" currency on the other, is all about probabilities. The chances of snaring all six numbers (1 in 13,983,816) do not diverge from the chances of snaring five (1 in 55,492) by the same degree that five diverges from four (1 in 1,033). Not by a long way. And so, whereas on one level things progress predictably, what they "boil down to," in a parallel mathematical universe, doesn't. What they map onto takes on a life of its own.

Back at the restaurant, I put my theory to Scott Lilienfeld: that actually both he and Joe Newman might be right. Psychopathy might well be on a spectrum. But at the sharp psychopathic business end of it, something ineffably stark seems to happen. A switch just seems to flip.

"I certainly think that's one way of reconciling the two perspectives," he reflects. "And it's undoubtedly the case that those at the extreme end of many distributions seem to run on a different kind of gas from everyone else. But it also depends on your starting point: whether you view psychopathy predominantly as a personality predisposition or as an information-processing disorder. Whether you want to deal in cognitive deficits or variations in temperament. You can see it in the language, in the terminology used: disorder; deficit; predisposition; variation . . . It would be interesting to hear what Joe has to say. Have you put it to him?"

I hadn't. But not long afterward, I did. "Is it possible," I asked Newman, "that the further along the psychopathic spectrum one gets—assuming such a thing exists—the more you start, neurologically speaking, to see gradual changes occurring? Say, differences in the brain's attentional mechanisms or reward systems, which, the more psychopathic an individual is, become increasingly laser-like in their focus, increasingly primed for immediate gratification? And that although performance on the PPI or PCL-R may be linear, the way that performance manifests itself in low-level brain activity, especially at very high scores, might instead be rather different? That it might, in fact, be spectacularly exponential?"

His eyes narrowed. The wily old gunslinger was in no mood for games. "Sure," he said. "It's possible. But the clinical cutoff [on the PCL-R] is 30. And that, in the lab, coincidentally or otherwise, is also the point at which most of the empirical shit hits the low-level cognitive fan."

He smiled and poured some coffee. "At any rate," he said, "it doesn't really matter which way up you hold it. A clinical psychopath is a pretty distinct specimen. Either way they're different. Right?"

THREE

CARPE NOCTEM

I have given suck, and know
How tender 'tis to love the babe that milks me:
I would, while it was smiling in my face,
Have pluck'd my nipple from his boneless gums,
And dash'd the brains out, had I so sworn as you
Have done to this.

—LADY MACBETH (on hearing that her husband plans to proceed no further with the murder of King Duncan)

The Devil and the Deep Blue Sea

On March 13, 1841, the *William Brown* set sail from Liverpool, bound for Philadelphia. Five weeks into the voyage, on the night of April 19, the vessel struck an iceberg 250 miles off the coast of Newfoundland and began to sink rapidly. Over thirty passengers and crew, still dressed in their nightclothes, commandeered a longboat built to hold just seven. With a storm looming and icy Atlantic rain already beginning to fall, it soon became apparent, to First Mate Francis Rhodes, that the longboat would have to be lightened were anyone to survive. The same thought had occurred to the captain, George L. Harris, who had taken to an accompanying jolly boat along with a handful of others. But he prayed for deliverance from another, more palatable source.

"I know what you'll have to do," he confided to Rhodes. "Don't speak of that now. Let it be a last resort." The following morning he sailed for Nova Scotia, leaving the hapless, foundering longboat to its fate.

On the day of the twentieth, and on into the night, conditions worsened and the waves began to build. The boat sprang a leak and, despite frantic bailing, began to take on water. The situation was hopeless. And so at ten o'clock, on the night of April 20, a momentous decision was taken: some individuals would have to be put to the sword. Such an action, reasoned Rhodes, would not be unjust to those who went over the side, for they would surely have perished anyway. But if, on the other hand, he deigned to take no action, he would be responsible for the deaths of those whom he could have saved.

Unsurprisingly, not all of the assembly concurred with Rhodes's conclusions. The dissenters contended that if no action was taken and everyone drowned as a result, then no one would be responsible for the deaths. In contrast, they argued, if he endeavored to save some of the party at the expense of even one other, he could only do so through the active taking of life, and would end his days, and quite possibly everyone else's, as a murderer. That, by far, would constitute the greater evil.

Unmoved by the charge, Rhodes stuck to his guns. Since their only hope of survival depended upon staying afloat, not to mention, in addition, a Herculean feat of oarsmanship, the situation as it stood was untenable, he countered. Something, or someone, had to give. "Help me, God! Men, go to work!" Rhodes cried out to the deckhands as he and fellow crew member Alexander Holmes set about the grisly task of casting people off into the tumultuous, inky cauldron of the North Atlantic. At first the other seamen did nothing, prompting a second exhortation from Rhodes:

"Men! You must go to work or we shall all perish!"

The death tally started to rise. All fourteen of the male passengers were sacrificed, including two that were found to be hiding. Those who remained were just two married men and a boy, plus all but two of the women: the sisters of a man who'd previously gone over the side, and who chose, voluntarily, to join him.

Eventually salvation beckoned and the survivors were rescued by a trawler bound for Le Havre. And when, at last, they arrived in Philadelphia, they filed a lawsuit with the district attorney. On April 13, 1842, almost a year to the day since he'd cheated the icy Atlantic, able seaman Alexander Holmes stood trial on a charge of murder. He was the only crew member they could find in Philadelphia—and was the only one to ever be indicted for his actions.

If you had been on the jury, how would you have viewed the case?

Before you answer, let me tell you why I'm asking. A couple of years ago, I presented this dilemma to a group of male undergraduates, half of whom scored high on the PPI and half of whom scored low. Each was given three minutes' deliberation time in which to mull the problem over; then each student submitted his verdict, anonymously, in a sealed envelope. I wanted to know if the difference in PPI scores would have a bearing on what they decided. It didn't take me long to find out.

Of the twenty volunteers who scored low on the PPI, only one reached a verdict within the allotted time. The others were still deliberating. But of the twenty volunteers at the other end of the scale, it was a different story entirely. Without exception, all of them had made their minds up, and the results were unanimous. Holmes was free to walk.

Thinking Outside the Group

If you're trying to get your bearings inside this ethical hall of mirrors, don't panic. The good news is that you're obviously not a psychopath. In actual fact, on April 23, 1842, ten days after the trial first opened, it took sixteen hours for the jury to return a verdict—almost as long as Holmes had spent in the water. Guilty he might have been—of manslaughter, that is, not murder—but under such psychological g-force that right and wrong had imploded under the pressure, becoming morally indistinguishable from each other. The judge handed Holmes a token six-month sentence, plus a twenty-dollar fine.*

In contrast, consider the following case as reported in the *Daily Telegraph* back in 2007:

> Two Police Community Support Officers did not intervene to stop a 10-year-old boy from drowning because they were "not trained" to deal with the incident, a senior police officer said today. The [officers] stood at the edge of a pond at a Wigan beauty spot as Jordon Lyon got into trouble while trying to rescue his eight-year-old step-sister. Two fishermen in their 60s jumped in and managed to save the girl, but the officers, who arrived at the scene shortly afterwards, did not attempt a rescue, deciding to wait until trained officers arrived. At the inquest into his death today the boy's distraught parents demanded to know why more effort was not made to save their son. [His] stepfather said: ". . . You don't have to be trained to jump in after a drowning child."

At first glance, this case and that of able seaman Alexander Holmes have little in common. In fact, they appear to be polar opposites. The former revolves around an extraordinary reluctance to preserve life; the latter, around a curious ambivalence toward saving it. Yet look a little closer and striking similarities emerge. In both scenarios, for instance, it's the breaking of rules that's the problem. In the Jordon Lyon affair, the officers were paralyzed by a code of conduct: an all-consuming requirement to toe the party line. Like performing seals, they had been trained beyond their instincts. Trained, you might say, to eschew any action for which they had *not* been trained. In the *William Brown* tragedy, the "rules" were more deeply encoded—they

* The Alexander Holmes verdict held that seamen have a duty to their passengers that is superior even to their own lives. In addition, it stipulated that the traditional defense of self-preservation was not always sufficient in a murder trial if the accused was under a special obligation to the deceased.

were more functional, and more "ethically hygienic." Yet they were, one could argue—as some quite vehemently did—no less detrimental to the exigencies of the moment. The seamen, so to speak, were in exactly the same boat as the police officers. Caught in the moral crosshairs on a bleak humanitarian knife-edge, they had to act quickly, decisively, and with manifest disregard for the consequences of their actions. Some did it better than others.

Yet alongside the challenge to our existential comfort zones, these two accounts also conceal, deep within the lining of their tragedies, a rather odd paradox. The fact that conformity is built into our brains is about as nailed down an evolutionary certainty as you can get. When a herd animal is threatened by a predator, what does it do? It huddles closer to the group. As individual salience decreases, chances of survival increase. This is just as true in humans as it is in other species. Streaming behind our supersonic, turbocharged brains are ancient Darwinian vapor trails stretching all the way back to the brutal, blood-soaked killing fields of prehistory. In an experiment, for instance, that hitched the latest in social networking to its earliest biological origins, social psychologist Vladas Griskevicius, then at Arizona State University, and his coworkers found that when users of an Internet chat room are made to feel under threat, they show signs of "sticking together." Their views display convergence, and they become more likely to conform to the attitudes and opinions of others in the forum.

But there are clearly times when the opposite is true: when the ability to break free of social convention, to "think outside the group," can also be a lifesaver—both literally and metaphorically. In 1952, the sociologist William H. Whyte coined the term "groupthink" to conceptualize the mechanism by which tightly knit groups, cut off from outside influence, rapidly converge on normatively "correct" positions, becoming, as they do so, institutionally impervious to criticism: indifferent to out-group opposition, averse to in-group dissent, and ever more confident of their own unimpeachable rectitude. The psychologist Irving Janis, who conducted much of the empirical work on the phenomenon, describes the process as "a mode of thinking that people engage in when they are deeply involved in a cohesive in-group, when the members' strivings for unanimity override their motivation to realistically appraise alternative courses of action." It's not exactly conducive to good decision making.

As a case in point, take the space shuttle *Challenger* fiasco. Under considerable political pressure to get things under way (Congress, at the time, was seeking a large slice of revenue in furtherance of the space program, and a series of problems had already delayed the launch), scientists and engineers at NASA appeared systemically immune to concerns raised by a coworker, just twenty-four hours before liftoff, over the O-rings in the booster rockets. Though a string of conference calls had specifically been convened to discuss the problem in detail, the decision, incomprehensible in hindsight, was made to press on. The goal, after all, was to get the show on the road.

In the event, it proved disastrous. Inquests revealed, as the villains of the piece, not just the O-rings, but another, more viral, more insidiously carcinogenic culprit: a musty, asphyxiating psychology. The Rogers Commission, a dedicated task force set up by then President Ronald Reagan to investigate the accident, confirmed the nagging, unspoken fears of social psychologists the world over: that NASA's organizational culture and decision-making processes had played a significant role in the lead-up to the tragedy. Pressure to conform, discounted warnings, sense of invulnerability. It was all there, plain as day.*

So is the capacity to stand alone, to play by one's own rules outside the normative safe haven of society, also hardwired? There's evidence

* The complete inventory of groupthink symptoms runs as follows: feelings of invulnerability creating excessive optimism and encouraging risk taking; discounting of warnings that might challenge assumptions; unquestioned belief in the group's morality, causing members to ignore the consequences of their actions; stereotyped views of enemy leaders; pressure to conform against members of the group who disagree; shutting down of ideas that deviate from the apparent group consensus; illusion of unanimity; "mindguards"—self-appointed members who shield the group from dissenting opinions (Janis, 1972).

to suggest that it is—and that a fearless, untroubled minority has evolved within our midst.

The Mathematics of Madness

Just how psychopathy got a toehold in the gene pool is an interesting question. If the "disorder" is so maladaptive, then why does its incidence remain stable across time, with an estimated 1 to 2 percent of the population qualifying as psychopathic? Andrew Colman, professor of psychology at the University of Leicester, has an equally intriguing answer—one, I suspect, that will forever be close to my heart after a recent entanglement with the Newark Airport interchange.

In 1955, the film *Rebel Without a Cause* made its cinematic debut. Never before had rebellious, misunderstood youth been portrayed so sympathetically on the silver screen. But enough of the armchair criticism. For game theorists at least, one scene towers head and shoulders above the rest: the one in which Jim Stark (played by James Dean) and Buzz Gunderson (played by Corey Allen) hurtle, in a pair of stolen cars, inexorably toward the edge of the cliff in a deadly game of chicken.

Let's think about that scene for a moment from the point of view of the drivers, says Colman. Or rather, think about a more familiar version of it in which the two protagonists accelerate directly toward each other in an impending head-on collision. Each of them has a choice: adopt the sensible, "non-psychopathic" strategy of swerving to avoid a pile-up, or choose the risky "psychopathic" one of keeping their foot on the gas. These choices, with their differential "payoff points," constitute a classic you-scratch-my-back-I'll-scratch-yours-or-then-again-maybe-I-won't scenario that we can model using game theory—a branch of applied mathematics that seeks to quantify optimal decision-making processes in situations where outcomes depend not on the actions of the individual parties involved, but rather on their interaction (see figure 3.1).

	(BUZZ) NON-PSYCHOPATHIC	(BUZZ) PSYCHOPATHIC
(JIM) NON-PSYCHOPATHIC	Jim gets 3 points Buzz gets 3 points	Jim gets 2 points Buzz gets 4 points
(JIM) PSYCHOPATHIC	Jim gets 4 points Buzz gets 2 points	Jim gets 1 point Buzz gets 1 point

Figure 3.1. A game-theoretical model of the evolution of psychopathy

If Jim and Buzz both go for the sensible option and swerve away from each other, the outcome is a draw with second-best payoffs going to each (3). In contrast, if both are psychopathic and decide to see it through, each risks death—or, at best, serious injury. And thus each receives the very worst payoff (1).

As Colman explains, however, if one driver—let's say Jim—opts for caution, while Buzz turns out to be "nuts," a differential suddenly appears. Jim drops points and gets the "chicken" payoff (2), while Buzz lucks out, with a maximum haul (4).

It's a mathematical microcosm of what rubbing shoulders with psychopaths (and the Newark Airport interchange) is actually like. And biologically it works: when the game is played repeatedly in the lab, by computer programs specifically encoded with predetermined response strategies, something very interesting happens. When the payoffs are converted into units of Darwinian fitness and the assumption is made that those players in receipt of larger payoffs give rise to a greater number of offspring who then adopt precisely the same strategy as their progenitors, the population evolves to a stable equilibrium in which the proportion of individuals consistently behaving psychopathically actually mirrors the observed incidence of the disorder in real life (around 1 to 2 percent).

Whoever keeps their foot on the gas—whoever keeps their nerve—is always going to win: provided, that is, that their opposite number is sane. Behaving "irrationally" might actually sometimes be rational.

In 2010, Hideki Ohira, a psychologist at Nagoya University, and his doctoral student Takahiro Osumi validated Colman's theory for real. Psychopaths, they discovered, under certain extraordinary circumstances make better financial decisions than the rest of us, for precisely the reason that Colman had so elegantly demonstrated. They behave in a manner that would otherwise appear irrational.

To demonstrate, Ohira and Osumi deployed the ultimatum game—a paradigm widely used in the field of neuroeconomics, which explores, broadly speaking, the way we evaluate primarily monetary, but also certain other types of gain. The game involves two players interacting to decide how a sum of money they are given should be divided. The first player proposes a solution. The second player decides whether or not to accept the offer. If the second player decides to reject it, then both of the protagonists get nothing. But if the second player decides to accept, then the sum is split accordingly.

Take a look at figure 3.2 and you'll notice something interesting about the game. The offer that player 1 puts on the table can either be fair or unfair. They can propose to split the money, say, 50-50. Or alternatively, say, 80–20. Now, usually what happens is this. As proposals start approaching the 70-30 mark (in favor of player 1), player 2 goes into rejection mode.* After all, it's not just about the money. There's a principle at stake here, too!

* Previous research has shown that offers below 20 to 30 percent of the stake have a roughly 50 percent chance of being rejected. (See Werner Güth, Rolf Schmittberger, and Bernd Schwarze, "An Experimental Analysis of Ultimatum Bargaining," *Journal of Economic Behavior and Organization* 3, no. 4 (1982): 367–88.)

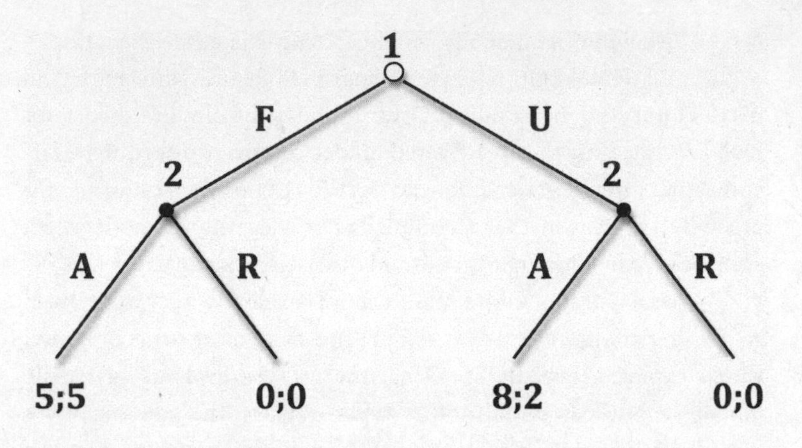

Figure 3.2. The ultimatum game (1 = Player 1; 2 = Player 2; F = fair; U = unfair; A = accept; R = reject)

But psychopaths, Ohira and Osumi discovered, play the game rather differently. Not only do they show greater willingness to accept unfair offers, favoring simple economic utility over the exigencies of punishment and ego preservation, they are much less bothered by inequity. On measures of electrodermal activity (a reliable index of stress based on the autonomic response of our sweat glands), the difference between psychopaths and other volunteers was telling, to say the least. Psychopaths were far less fazed than controls when screwed by their opposite numbers—and at the conclusion of the study, had more in the bank to show for it. A thicker skin had earned them thicker wallets.

Sometimes, Ohira and Osumi concluded, it pays to be a psychopath—but in a different way to that shown by Andrew Colman. Whereas Colman had demonstrated that it was good to put the boot in (or in his case, put it down), Ohira and Osumi had discovered the complete reverse. There was similar worth in taking it on the chin.

If you need any convincing of the value of either strategy, just ask someone who's been in the can.

To Get to the Top, Send Your Reputation Up Ahead of You

"Like a flashy, violent streak across the prison sky" is how one private investigator has described them. And there aren't too many, on either side of the bars, who would disagree with him. The Aryan Brotherhood, also known as the Rock, is one of the most feared gangs ever to emerge within the U.S. federal penitentiary system. Responsible, according to FBI figures, for 21 percent of murders inside U.S. prisons (though their members account for a mere 1 percent of inmates), you can't exactly miss them. Members display walrus-like mustaches more befitting the Wild West than a modern-day outlaw, and tattoos depicting a shamrock fused with a swastika, with the motif "666" emblazoned upon its leaves. Sport one without permission and you're invariably asked to remove it. Usually with a razor.

Brutally elite, the Rock is the Special Forces of the prison world. Founded in California's San Quentin "Supermax" high-security facility in 1964 by a group of white supremacists, the Brotherhood was numerically smaller than other prison gangs, but within a matter of just a few blood-spattered months had skyrocketed to top-dog status. How did they manage it? Well, it doesn't hurt to be smart, that's for sure. Despite the fact that many gang members were incarcerated in other Supermax units, often under conditions of twenty-three-hour lockdown, they managed to coordinate their activities through a number of ingenious methods: invisible ink made from urine and a four-hundred-year-old binary code system devised by the Renaissance philosopher Sir Francis Bacon, no less, being a couple of notable examples.

But they were also utterly remorseless, and lived (as still they do today) by one simple, sinister code: "Blood in, blood out." Blood in: every prospective member is admitted on the basis of their having already killed a member of a rival gang, and on the understanding that they will carry out further executions to order. Blood out: their only exit card is their own, often hastened, demise, whether through an event as vanishingly improbable as natural causes or, as is infinitely more likely (and, in many cases, more preferable), through similarly violent means.

As members admit, it's a mercilessly minimalist philosophy. There are no half measures and no questions asked. "Fear nothing and no one" is the mantra. And what the Rock lacks in numbers, it makes up for with nerveless ferocity. Not to mention, as is common in highly motivated psychopaths, ruthless dedication to the task.

With access to prison libraries (plus supplementary reading materials from other, less official sources), members treat killing like an undergrad science module, poring over human anatomy texts (alongside Nietzsche, Machiavelli, Tolkien, and Hitler) to find the parts of the body most vulnerable to sudden trauma. In the warped space-time continuum that exists within a Supermax prison, a ten-second window is like a wormhole into eternity—and a fight of such magnitude on the inside equates to a twelve-round slugfest in the extended, relativistic orbit of everyday life. Speed is of the essence. In the blink of an eye, much can be accomplished: Windpipes severed. Jugulars ripped out. Spinal cords pierced. Spleens and livers punctured. It's important to know what you're doing should the opportunity present itself.

Yet as Barry, one former member of the Rock, pointed out to me, in the impenetrable moral crevices that lurk, unseen and ungovernable, in the fear-darkened corners of a federal penitentiary, such a strategy might actually be construed as adaptive—as firefighting instead of fire-setting. And might, in the long run, contain trouble instead of igniting it.

"Prison," elucidates Barry, "is a hostile environment. It has a different set of rules than the outside world. It's a community within a community. If you don't stand up and be counted, someone can move in on you any time they want. So you have to do something about it. You don't have to keep taking people out. That ain't the way it works. Once or twice is usually enough. You do it once or twice and word soon gets round: Don't mess with these guys. Prevention, is what I'm saying, is better than cure. *Carpe noctem*."

Barry's point about conflict resolution is an interesting one, and is echoed, in not so many words, by the incarcerated record producer Phil Spector. "Better to have a gun and not need it," the Magnum-toting screwball once expounded, "than to need a gun and not have it" (though whether he still believes that today is anybody's guess). A more nuanced position is taken by the Chinese military strategist of the sixth century B.C., Sun Tzu. "To subdue the enemy without fighting," wrote Sun, "is the highest skill"—a skill, as we saw just a moment or so ago with Jim and Buzz, that's both hard to fake and unequivocally rooted in confidence. Not a false confidence based on bravado. But a real confidence based on belief.

Here's Dean Petersen, an ex–Special Forces soldier turned martial arts instructor: "Sometimes, when you're in a hostile situation, your best option is to match the aggressive intentions of a potentially violent individual. And then go one step beyond them. Raise them, in other words, to use a poker analogy. Only then, once you've gained the psychological ascendancy, shown them . . . hinted . . . who's boss, can you begin to talk them down."

How better to assert your authority than by convincing prospective challengers that they're beaten before they start?

Barry's argument has wider implications, too—for the selection, not just of ruthlessness, but of other psychopathic characteristics such as fearlessness and superficial charm. Conflict, it transpires, isn't the only means of establishing dominance in the natural world. Back in the days of our ancestors, survival, just as in prison, didn't come cheap. Although group membership constituted a significant chunk of the price tag, communities also placed a surprisingly high premium on risk takers.

One observes a similar dynamic in monkeys still today. Male chimpanzees (our closest living relative, with whom we share 96 percent of our DNA) will compete through "magnanimity": through the direction of unsolicited altruism toward subordinates. Such magnanimity is usually gastronomic in nature: enduring danger to provide the troop with food, sharing out the proceeds of one's own kills charitably, and confiscating those of others for the purposes of reallocation.

As the primatologist Frans de Waal points out, "Instead of dominants standing out because of what they take, they affirm their position by what they give."

Of equal note are those primates who vie with one another for status through "public service" or "leadership"—by facilitating cooperation within the group, or, if you prefer, through charisma, persuasion, and charm. Dominant chimpanzees, stump-tailed monkeys, and gorillas all compete by intervening in disputes among subordinates. Yet, contrary to expectation, such intervention does not, by default, automatically favor family and friends. It is implemented, as de Waal observes, "on the basis of how best to restore peace."

Consequently, de Waal continues, rather than decentralizing conflict resolution, "the group looks for the most effective arbitrator in its midst, then throws its weight behind this individual to give him a broad base of support for guaranteeing peace and order."

Ruthlessness. Fearlessness. Persuasiveness. Charm. A deadly combination—yet also, at times, a lifesaving one. Have the killers of today enjoyed a sneaky evolutionary piggyback on the prowess of yesterday's peacemakers? It may not be beyond the bounds of possibility—though violence, of course, isn't exactly new.

The First Psychopaths

In 1979, at a remote site near the village of Saint-Césaire in southwest France, Christoph Zollikofer of the University of Zurich and a joint contingent of French and Italian researchers made an intriguing discovery. Dating back to the "transitional period," when prognathous-jawed, ridge-browed Europeans were undergoing displacement by an anatomically modern influx from Africa, the remains of a skeleton some 36,000 years old had lain in an anthropological coma since the Ice Age. The remains, it was confirmed, were Neanderthal. But there was something rather odd about the skull. It was scarred. The scar in question was on a section of bone approximately four centimeters in length, and was situated top right. It was not, of course, unheard of for excavations in the field to throw up less than perfect specimens. In fact, it was to be expected. But there was something, somehow, just a little different about this one.

It had the whiff of premeditation about it that suggested foul play; that alluded less to the vicissitudes of geophysical atrophy and more to the exigency of a prehistoric moment, lost deep in the lining of our dark ancestral past. This was no ordinary tale of misadventure, but a lesion caused by violence. Or, more specifically, by a slashing or hacking motion indicative of a sharp-bladed implement. Putting two and two together—the position of the scar; the shape of the wound; the fact that the rest of the skull appeared neither fractured nor misshapen—Zollikofer arrived at a stark conclusion. Interpersonal aggression among humans had a longer lineage than had previously been suspected. Inflicting harm on others came, it would seem, quite naturally.

It's an intriguing thought that itinerant Neanderthal psychopaths were doing the rounds of prehistoric Europe some 40,000 years ago. But it's not all that surprising. Indeed, in contrast to the "piggyback" argument just outlined, the traditional take on the evolution of psychopathy focuses, as we saw in the previous chapter, predominantly on the predatory and aggressive aspects of the disorder. On one of the standard psychopathy assessment questionnaires, the Levenson Self-Report Scale, a typical test item reads as follows:

"Success is based on survival of the fittest. I am not concerned about the losers." On a scale of 1 to 4, where 1 represents "strongly disagree" and 4 represents "strongly agree," rate how you feel about this statement.

Most psychopaths are inclined to register strong agreement with such a sentiment—which is not, incidentally, always a bad thing.

"Two little mice fell in a bucket of cream," says Leonardo DiCaprio, playing the role of Frank Abagnale, one of the world's most celebrated con men, in the film *Catch Me If You Can*. "The first mouse quickly gave up and drowned. The second mouse wouldn't quit. He struggled so hard that eventually he churned that cream into butter and crawled out . . . I am that second mouse."*

Yet, at the other end of the spectrum, we run into an altogether different kind of exhortation, such as those espoused in religious, spiritual, and philosophical texts. We find allusions to temperance, tolerance, and the meek inheriting the earth.

So which one are you: psychopath, saint, or somewhere in between? The chances are it's going to be the latter—for which, it turns out, there are sound biological reasons.

To Plea or Not to Plea

We've already seen game theory in action earlier in this chapter. A branch of applied mathematics devoted to the study of strategic situations, to the selection of optimal behavioral strategies in circumstances in which the costs and benefits of a particular choice or decision are not set in stone but are, in contrast, variable, game theory presents scenarios that are intrinsically dynamic. Unsurprisingly perhaps, given game theory's inherent emphasis on the relationship between individual agency and the wider social group, it's not uncommon to find rich incrustations of this semiprecious mathematical outcrop embedded within branches of natural selection—within models and theories of how various behaviors or life strategies might have evolved. Psychopathy, as the work of Andrew Colman has shown us, is no exception.

In order to take up where Colman left off and explore the evolutionary dynamics of the psychopathic personality further, let's rig up a situation similar to the one Jim and Buzz found themselves in on

*I haven't assessed Frank Abagnale, but in his prime, he certainly seemed to demonstrate many of the hallmarks of the psychopath. It matters not. Even if I had assessed him, he would probably have succeeded in faking his test results anyway!

the cliffs—only this time, make it a little more personal. Imagine that you and an accomplice are suspected of committing a major crime. The police picked you up and have taken you in for questioning.

Down at the station, the chief investigating officer interviews you both separately—but has insufficient evidence to press charges, so he resorts to the age-old tactic of playing one against the other. He puts his cards on the table and cuts you a deal. If you confess, he will use your confession as evidence against your partner and send him down for ten years. The charges against you, however, will be dropped and you will be allowed to walk away with no further action being taken. Too good to be true? It is. There's a catch. The officer informs you that he will also be offering the same deal to your partner.

You are left alone to ponder the arrangement. But during that time you suddenly have an idea. What if both of us confess? you ask. What happens then? Do we both go to prison for ten years? Or are both of us free to leave? The officer smiles. If both of you confess, he replies, he will send each of you to prison—but on a reduced sentence of five years. And if neither confesses? Prison again, but this time for only a year. (See figure 3.3.)

	PARTNER DOES NOT CONFESS	PARTNER CONFESSES
YOU DO NOT CONFESS	Partner gets 1 year You get 1 year	Partner goes free You get 10 years
YOU CONFESS	Partner gets 10 years You go free	Partner gets 5 years You get 5 years

Figure 3.3. The Prisoner's Dilemma

This officer is smart. Think about it. He has, in effect, made you an offer you can't refuse. The truth of the matter is simple. Whatever your partner may choose to do, you are always better off confessing. If your partner decides to keep his mouth shut, then you either face a year in the slammer for doing the same thing or stroll off into the sunset by informing against him. Similarly, if your partner decides to inform against you, then you either go down for the full term for deciding to hold out or halve your sentence by mirroring their betrayal. The reality of both of your predicaments is freakishly paradoxical. Logically speaking, self-preservation dictates that the only sensible course of action is to confess. And yet, it's this same paralyzing logic that robs you both of the chance of minimizing your joint punishment by remaining silent.

And note that the question of probity—remaining tight-lipped because it's the "right" thing to do—doesn't come into it. Quite apart from the dubious moral worth of placing oneself in a position that is self-evidently prone to exploitation, the whole purpose of the Prisoner's Dilemma is to ascertain optimal behavioral strategies not within frameworks of morality, with philosophical enforcers working the doors, but within a psychological vacuum of zero moral gravity . . . such as that which comprises the natural world at large.

So could the psychopaths be right? Could it really be survival of the fittest out there? Such a strategy, it would seem, is certainly logical. In a one-off encounter such as the Prisoner's Dilemma, you might argue that dog-eat-dog (or a strategy of defection, to use the official terminology) constitutes a winning hand. So why not, in that case, just go ahead and play it?

The reason, of course, is simple. Life, in its infinite complexity, doesn't go in for one-offs. If it did, and the sum total of human existence was an endless succession of ships passing in the night, then yes, the psychopaths among us would indeed be right, and would quickly inherit the earth. But it isn't. And they won't. Instead, the screen of life is densely populated with millions upon millions of individual pixels, the repeated interaction of which, the relationships between which, gives rise to the bigger picture. We have histories—social histories—with each other. And we are able, unlike the characters in the Prisoner's Dilemma, to communicate. What a difference that would have made! But that's okay—just as we are able to play the Prisoner's Dilemma the one time, so we can play it a number of times. Over and over. Substituting prison terms for a system of reward and punishment in which points are won or lost (see figure 3.4), we are able, with the aid of some simple mathematics, to simulate the complexity of real life, in exactly the same way as we did with Jim and Buzz.

	PARTNER COOPERATES	PARTNER COMPETES
YOU COOPERATE	Partner earns 5 points You earn 5 points	Partner earns 10 points You earn 0 points
YOU COMPETE	Partner earns 0 points You earn 10 points	Partner earns 1 point You earn 1 point

Figure 3.4. A sample Prisoner's Dilemma game

What happens then? Do the psychopaths cut it in a world of repeated encounters? Or is their strategy trumped by simple "safety in numbers"?

Saints Against Shysters

To answer this question, let's imagine a society slightly different from the one we currently live in: a society like that of days gone by, in which the workforce is paid in cash at the end of each week, in personalized little brown envelopes. Now imagine that we can divide this workforce into two different types of people. The first type is honest and hardworking and puts in a full week's work. Let's call them the saints. The other is dishonest and lazy and preys upon its diligent counterparts as they make their way home on a Friday, lying in wait outside the factory gates and appropriating their hard-earned wages for themselves. Let's call them the shysters.*

* A similar dynamic exists for real in apiculture. During times of scarcity, so-called robber bees will attack the hives of other bees, killing all within their path,

At first it would seem as if the shysters have got it made: that crime pays. And, indeed, in the short term at least, it does. The saints clock in to keep the community going, while the shysters reap a two-fold benefit. Not only do they enjoy the advantages of living in a flourishing society, they also, by stealing the saints' wages, "get paid" for doing nothing. Nice work if you can get it. But notice what happens if the pattern of behavior continues. The saints begin to tire and fall sick. Having less disposable income with which to look after themselves, they begin to die out. Gradually, the ratio of the "working" population starts to shift in favor of the shysters.

But this, of course, is precisely what the shysters don't want! With the number of saints diminishing by the week, the likelihood increases that the shysters will encounter each other. Moreover, even if they do run into a saint, there's a greater chance that they'll come away empty-handed. Another shyster may well have beaten them to it.

Eventually, if the fun and games are allowed to play out naturally, the power balance comes full circle. The pendulum swings back in favor of the saints, and society reverts to working for a living. But note how history is programmed to repeat itself. The saints call the shots for only such time as the economy is in recession, and the shysters preside for only as long as the saints can keep them afloat. It's a bleak carousel of recurring boom and bust.

This brief sketch of two very different work ethics is, to say the

---

including, in some cases, the queen, in order to appropriate their honey. Hives protect themselves from robbers by appointing guard bees on sentry duty at the hive entrance, to be on the lookout for the raiders and to fight them to the death in the event of an attack. In a study just out, however, a combined team of researchers from the University of Sussex in the U.K. and the University of São Paulo in Brazil has just uncovered the world's very first "soldier" bee. This subcaste of the Jatai bee (*Tetragonisca angustula*), unlike normal guard bees in honeybee colonies, is physically specialized to perform the task of protecting the hive. It is 30 percent heavier than its forager nest mates, and has larger legs and a smaller head. Perhaps they should call it the "berserker bee." (See Christoph Grüter, Cristiano Menezes, Vera L. Imperatriz-Fonseca, and Francis L. W. Ratnieks, "A Morphologically Specialized Soldier Caste Improves Colony Defense in a Neotropical Eusocial Bee," *PNAS* 109, no. 4 (2012):1182–86. doi:10.1073/pnas.1113398109.)

least, a simplified representation of an infinitely more complex set of dynamics. Yet it is precisely this simplification, this behavioral polarization, which lends such a model its power. Pure unconditional aggression and pure unconditional capitulation are destined to fail as strategies of social exchange in a society of multiple interaction and mutual dependence. In what essentially amounts to a peripatetic seesaw effect, each strategy is vulnerable to exploitation by the other once one has gained the ascendancy: once the proponents of one strategy become enough of a mob to be parasitized by those of the competing strategy. To coin a phrase from the sociobiology lexicon: as strategies for survival, neither unqualified cooperation nor unqualified competition may be regarded as evolutionarily stable.* Both may be trumped by invading or mutating counterstrategies.

But can we actually observe this iterative process in action, this repeated unfolding of the Prisoner's Dilemma dynamic? We are, after all, firmly in the realm of a thought experiment here. Do these abstract observations pan out in real life? The answer depends on what we mean by "real." If in "real" we're prepared to include the "virtual," then it turns out we might be in luck.

Virtual Morality

Suppose I were conducting an experiment on people's responses to the unexpected and I presented you with the following opportunity: For a thousand dollars, you must take off all your clothes and walk, stark naked, into a bar to join a group of friends. You must sit at a table and talk to them for five minutes (that's two hundred dollars a minute!), during which time you will feel the full force of the excruciating social embarrassment that will undoubtedly accompany the venture. However, after the five minutes have elapsed, you will leave the bar unscathed, and I will ensure that neither you nor anyone else

*This term was first introduced by the late John Maynard Smith of the Centre for the Study of Evolution, University of Sussex.

who was present will have any memory of the event. I shall erase it all. Apart from the crisp bundle of notes burning a hole in your pocket, it will be as if nothing had ever happened.

Would you do it? In fact, how do you know you haven't done it already?

There are some people, I'm sure, who would gladly bare all for the sake of scientific advancement. How liberating it would be if somehow, somewhere, in the terraces and tenements of time, we could check in and out of a transient, encapsulated world where experiences are rented by the hour. This, of course, was very much the theme of *The Matrix*: humans inhabiting a virtual world, which appeared, at the time, compulsively, compellingly real. But what of the flip side? What of computers inhabiting a world that is human?

In the late 1970s, the political scientist Robert Axelrod asked exactly this question in relation to the Prisoner's Dilemma—and hit upon a method of digitizing the paradigm, of determining a strategy, over time and repeated interaction, that ticked all the boxes of evolutionary stability. He sequenced the genome of everyday social exchange.

First, Axelrod approached a number of the world's leading game theorists about the idea of holding a Prisoner's Dilemma tournament in which the sole participants were computer programs. Second, he urged each theorist to submit a program to take part in the tournament that embodied a set, prespecified strategy of cooperative and competitive responses. Third, once all the submissions had been received (there were fourteen of them in all), he set up a preliminary round prior to the commencement of the contest's main event, in which each of the programs competed against the others for points. At the conclusion of this round, he added up the number of points that each program had accrued, and then kicked off the tournament proper, with the proportion of programs represented corresponding to the number of points that each had amassed in the preceding round—precisely in line with the strictures of natural selection. Then he sat back and watched what happened.

What happened was pretty straightforward. The most successful program by far was also, by far, the most simple. TIT FOR TAT, designed by the Russian-born mathematician and biologist Anatol Rapoport, whose pioneering work on social interaction and general systems theory has been applied to issues of conflict resolution and disarmament not just in the lab but on the political stage at large, did exactly what it said on the label. It began by cooperating, and then exactly mirrored its competitor's last response. If, on trial 1, for example, that competitor also cooperated, then TIT FOR TAT would continue to follow suit. If, on the other hand, the rival program competed, then on subsequent trials it got a taste of its own medicine . . . until such time as it switched to cooperation.

The graceful practicality and resilient elegance of TIT FOR TAT soon became apparent. It didn't take a genius to see what it was up to. It embodied, spookily, soullessly, in the absence of tissues and synapses, those fundamental attributes of gratitude, anger, and forgiveness that make us—us humans—who we are. It rewarded cooperation with cooperation—and then reaped the collective benefits. It took out immediate sanctions against incipient competition, thus avoiding the reputation of being a soft touch. And in the aftermath of such rancor, it was able to return, with zero recrimination, to a pattern of mutual back-scratching, nipping in the bud any inherent potential for protracted, destructive, retrospective bouts of sniping. Group selection, that hoary evolutionary chestnut that that which is good for the group is preserved in the individual, didn't come into it. If Axelrod's experiment showed us anything at all, it was this: altruism, though undoubtedly an ingredient of basic group cohesion, is perfectly capable of arising not out of some higher-order differential such as the good of the species or even the good of the tribe, but out of a survival differential existing purely between individuals.

Macroscopic harmony and microscopic individualism were, it emerged, two sides of the same evolutionary coin. The mystics had missed the point. Giving wasn't better than receiving. The truth, according to Robert Axelrod's radical new gospel of social informatics, was that giving *was* receiving. And, what's more, there was no known antidote. Unlike our earlier example of the saints and the shysters, in which a "tipping point" kicked in once the high end of the population seesaw assumed a certain level of ascendancy, TIT FOR TAT just kept on rolling. It was able, over time, to sweep all competing strategies off the field permanently. TIT FOR TAT wasn't just a winner. Winning was just for starters. Once it got going, it was pretty much invincible.

Best of Both Worlds

Axelrod's adventures in the world of "cybernethics" certainly raised a few eyebrows. Not just among biologists, but in philosophical circles, too. To demonstrate so convincingly that "goodness" was somehow inherent to the natural order, that it was an emergent property, as it were, of social interaction, succeeded only in driving an even bigger wedge between those on the side of God and those who put God to one side. What if our "better" nature wasn't better, after all? But was, instead . . . well, just nature?

Such an abomination had already occurred, a decade or so prior to Axelrod's endeavors, to a young Harvard biologist by the name of Robert Trivers, who somewhat presciently had speculated that perhaps it was for precisely this reason that certain human attributes had evolved in the first place: to spray-paint on the side of consciousness an affective affirmation of such a brilliantly simple blueprint, such a neat mathematical mantra, as TIT FOR TAT—a mantra that had undoubtedly served its apprenticeship in the ranks of the lower animals before we got our hands on it. Perhaps, Trivers mused, it was for this very reason that we experienced, for the first time in the depths of our evolutionary history, those initial flushes of friendship and enmity, of affection and dislike, of trust and betrayal, that now, millions of years on, make us who we are.

The seventeenth-century British philosopher Thomas Hobbes would almost certainly have approved. Some three hundred years earlier, in *Leviathan*, Hobbes had anticipated precisely such a notion with his concept of "force and fraud": the idea that violence and cunning constitute the primary, indeed the sole, instigators of outcomes. And that the only analgesic for "continual fear, and the danger of violent death; and the life of man, solitary, poor, nasty, brutish and short" is to be found in the sanctuary of agreement. The formation of alliances with others.

To be sure, the conditions of Axelrod's tournament certainly reflected those of human, and prehuman, evolution. Several dozen regularly interacting "individuals" was just about the right number as far as early communities went. Similarly, each program was endowed with the capacity not only to remember previous encounters, but also to adjust its behavior accordingly. So it was an intriguing notion, this theory of moral evolution. In fact, it was more than that. Given what had initially gone into Axelrod's mathematical sausage machine and what had come out the other end, it was an eminent possibility. "Survival of the fittest" now appeared not, as had been previously thought, to reward competition indiscriminately. But rather, to reward it discerningly. Under certain sets of circumstances, yes, aggression might open doors (one thinks of Jim and Buzz). But under others, in contrast, it might just as easily close them—as we saw with the saints and the shysters.

So the psychopaths, it transpires, have got it only half right. There's no denying the harshness of existence, the brutal, sawn-off truth that it can, at times, be survival of the fittest out there. But this is not to say that it has to be that way. The meek, it turns out, really do inherit the earth. It's just that along the way there are always going to be casualties. "Do unto others" has always been sound advice. But now, some two thousand years later, thanks to Robert Axelrod and Anatol Rapoport, we've finally got the math to prove it.

Of course, that there's a bit of the psychopath in all of us—a spectral biological fugitive from the algebra of peace and love—is beyond doubt, as it is that our overlords from the bureau of natural selection have granted psychopaths ongoing evolutionary asylum down the years. Sure, the moral of the saints and the shysters might be set in

Darwinian stone: If everyone floors it, there'll eventually be nobody left. But equally, there are times during the course of our everyday lives when we *all* need to pump the gas. When we all, rationally, legitimately, and in the interests of self-preservation, need to calmly "put our foot down."

Let's return to Axelrod's virtual free-for-all one last time. The reason that TIT FOR TAT rose to the top of the heap in such remorseless, unstoppable fashion was because beneath the smiley exterior lurked a hidden inner steel. When the situation demanded it, it wasn't in the least bit squeamish about putting its silicon foot on its rival's neck. Quite the reverse, in fact. It evened the score as soon as the opportunity presented itself. The secret of TIT FOR TAT's success lay as much in its ruthless dark side as it did in its default sunny side; in the fact that when the going got tough, it was able to step up to the plate and mix it up with the best of them.

The conclusions are as clear as perhaps they are unnerving. TIT FOR TAT's blueprint for success certainly has psychopathic elements to it. There's the surface charm on the one hand. And the ruthless quest for vengeance on the other. Then, of course, there's the nerveless self-assurance to return to normal as if nothing had ever happened. The program is no Aryan Brotherhood, that's for sure. But between the switches and the soulless synaptic twitches lurk echoes of their creed. Speak softly and carry a big stick, goes the phrase. Good advice, if you want to get ahead—in both the virtual and the real worlds. Which is why, to return to our question of earlier, psychopaths still walk the earth, and haven't sunk without trace beneath the deadly Darwinian currents that terrorize the gene pool.

There will always be a need for risk takers in society, as there will for rule-breakers and heartbreakers. If there weren't, ten-year-old boys would be falling into ponds and drowning all over the place. And who knows what would happen at sea? If First Mate Francis Rhodes and able seaman Alexander Holmes hadn't dredged up the courage to set about doing the unthinkable, one wonders if there would have been any survivors of that fateful night in 1841, 250 miles off the icy coast of Newfoundland, in the raging North Atlantic.

FOUR

THE WISDOM OF PSYCHOPATHS

Just because I don't care doesn't mean I don't understand.

—HOMER SIMPSON

New Year Resolution

My oldest friend is a psychopath. We go all the way back to nursery school. I remember one of the teachers taking me over to the sandbox and introducing me to this blond, roly-poly kid who was playing with one of those puzzles where you have to insert the right shape into the right hole. Anyway, I picked up a star, and tried to shove it through the hole that, with the benefit of hindsight, I can now clearly see was most definitely intended for the parrot. It wouldn't fit. Worse, it got stuck. Johnny spent twenty seconds or so (an eternity in the life of a five year-old) calmly working it free. And then he poked me in the eye with the damn thing. That callous, unprovoked, and, frankly, downright juvenile attack pretty much marked the high point of our friendship.

Fast-forward ten years or so and Johnny and I are in high school together. It's recess, and he comes over to me and asks if he can borrow my history paper. He's "left his at home." And guess which class is next? "Don't worry," says Johnny. "There'll be no way of telling. I'll make it look completely different."

I hand him the paper and catch up with him again at the beginning of class. "You got my paper, Johnny?" I whisper.

Johnny shakes his head. "Sorry," he says. "No can do."

I start to panic. This particular teacher isn't the kind you mess with—no paper would mean no grade. Plus detention.

"What do you mean, no can do?" I hiss. "Where is it?"

Calm as you like, as if he's narrating a bedtime story, Johnny spills the beans. "Well, Kev, it's like this," he explains. "You see, I didn't have time to rewrite it, like I'd said. So I copied it out verbatim."

"But," I shriek as the teacher, who's not exactly noted for his people skills, stomps into the classroom, "that doesn't explain where mine is, does it?"

Johnny looks at me as if I'm utterly insane. "Well, we couldn't both hand in the same piece of work, could we?" he says.

"No!" I exclaim, clearly still not getting it. "We couldn't! So where the hell's my paper?"

Johnny shrugs. And takes out "his" work for collection.

"It's in the trash," he says casually. "Behind the music building."

Instinctively, I spring out of my chair. Maybe there's time to retrieve it before the class kicks off.

"You asshole," I snarl under my breath. "I'm going to fucking kill you."

Johnny grabs my arm and yanks me back down by the sleeve. "Look," he says with a concerned, paternalistic smile, gesturing over at the window. "It's pissing rain out there, and you're going to get soaked. You don't want to ruin your chances of breaking that school mile record next week by coming down with something, do you?"

There's not a hit of irony in Johnny's tone. I've known him long enough to realize that, actually, he genuinely believes he's looking out for me. He really does think he's got my best interests at heart. Infuriatingly, in this instance, I have to agree with him. The bastard's got a point. The record has stood since the early sixties. And the training's been going well. Shame to ruin all the hard work by doing something stupid at the last minute.

I slump back down in my seat, resigned to my fate.

"Good man," says Johnny. "After all, it's only a paper. Life's too short."

I'm not listening. Already, I'm trying to come up with a plausible explanation as to why I don't have the piece to hand in. And how, if the rain damage isn't too extensive, I can dry it out—or, failing that, copy it out—and submit it later.

I don't have long to engineer my cover story. The Grim Reaper is already on his rounds, and is now only a couple of rows in front of us, a sententious pile of crap on the Franco-Prussian War burning a fulsome little hole in his clutches.

Johnny scoops up his contribution and casts an admiring eye over it. Then he pats me on the back and, glancing out the window, screws up his face at the rain.

"Besides," he adds, "you'd have been too late anyway, Kev. I guess I should qualify what I just said. What's left of it is in the trash. Actually, I burned it, mate."

You may be wondering why on earth I've remained friends with Johnny all these years. And sometimes, in my more reflective moments, I wonder the same thing myself. But don't forget that Johnny is a psychopath.* And, as we know, they often have saving graces. One of Johnny's is his uncanny ability to turn virtually any situation to his own advantage—not uncommon among highly intelligent members of his species. He is, without doubt, one of the most persuasive people I've ever known (and I include in that brotherhood a number of the world's top con artists). Not only that, but he is, I guess you could say, a persuasion prodigy.

When we were about five or six, Johnny's folks had to attend a funeral in Canada. Johnny stayed behind and spent New Year's Eve at my house. It got to around nine o'clock, and my parents started

*It was while we were at college that I gave him the PPI, the questionnaire, if you recall from the previous chapter, specifically developed by Scott Lilienfeld and Brian Andrews to assess psychopathic attributes, not in incarcerated offenders, but in the general population. Not surprisingly, he scored extremely high, in particular on Machiavellian Egocentricity, Carefree Nonplanfulness, Social Potency, Stress Immunity, Fearlessness, and Coldheartedness (six of the eight subscales that make up the questionnaire, the other two being Blame Externalization and Impulsive Nonconformity).

dropping hints that it was time to go to bed. Hints like "It's time to go to bed." Like any self-respecting six-year-old, I didn't take it lying down.

"But, Mum," I whined. "Johnny and I want to stay up until midnight. Please . . . !"

She wasn't having any of it. But this, needless to say, didn't stop me from coming up with a veritable catalogue of mitigating circumstances, ranging from the fact that all our friends were allowed to stay up late at New Year's (original, huh?) to the rather profound observation that the New Year does indeed only come round once a year. Johnny, however, remained conspicuous by his silence. He just sat there, as I recall, listening to the drama play out. Taking it all in like some top city defense lawyer waiting for his moment to pounce.

Finally, Mum had had enough. "Come on!" she said. "That's it! You know what you're like when you stay up late. You get cranky and irritable, and the next day you don't get out of bed until noon."

Reluctantly, despondently, and with a creeping sense of end-stage resignation, I looked across at Johnny. The game was up. It was time to say good night. But no one had bargained for what happened next. With perfect timing, just as I was about to throw in the towel and start heading upstairs, Johnny broke his silence.

"But, Mrs. Dutton," he said. "You don't want us running around at the crack of dawn tomorrow morning while you're lying in bed with a headache, do you?"

We went to bed at three.

The Dark Triad and James Bond Psychology

Johnny's ability to wheel and deal at life's flipping points, to make the absolute most out of whatever situation he found himself in, eventually stood him in good stead. He joined the secret service.

"It's not just the cream that rises to the top, Kev," he would say. "It's the scum, too. And you know what? I'm both. Depends on what takes my fancy." It's difficult to fault such coruscating insight.

Needless to say, the fact that Johnny went and got a job with MI5—the British equivalent of the FBI—didn't surprise any of us. And, whatever it is that he does for them, he is, by all accounts, pretty good at it. Such is his coolness, charisma, and demonic power of persuasion, one of his colleagues once told me at a party, that even if he had a telephone cord wrapped around your neck, he'd be charming the bloody pants off you.

"He'd strangle you with his own halo," the guy said. "And then put it back on as if nothing had ever happened." I didn't need any convincing.

Of course, if by this stage Johnny is starting to remind you a little bit of James Bond, it's no coincidence. It's easy to imagine how that other notable employee of Her Majesty's Secret Service might also be a psychopath; how the shadowy world of spooks, countersurveillance, and espionage might well be wall-to-wall with under-the-radar serial killers, with a license to kill rather than some deep, unfathomable compulsion; and how, were the Walther PPK–packing secret agent that we all know and love to swap that PPK for a copy of the PPI, he might be pretty high on the spectrum. But is there any basis behind such speculation? Buying into the stereotype is one thing; seeing how the fantasy plays out in reality, quite another. Is it pure, unadulterated chance that Johnny is a psychopath—and just so happens to work in the field of military intelligence?

One man who asked these questions, and then set about finding the answers, is psychologist Peter Jonason. Back in 2010, Jonason (then at New Mexico State University) and his colleagues published a paper titled "Who Is James Bond? The Dark Triad as an Agentic Social Style," in which they showed that men with a specific triumvirate of personality traits—the stratospheric self-esteem of narcissism; the fearlessness, ruthlessness, impulsivity, and thrill-seeking of psychopathy; and the deceitfulness and exploitativeness of Machiavellianism—can actually do pretty well for themselves out there in certain echelons of society. Not only that, but they're also, in addition, more likely to have a greater number of sexual partners and a stronger inclination toward casual, short-term relationships than men who are low on such traits. Far from the Dark Triad constituting a handicap when it comes to dealing with the opposite sex, contends Jonason, it may, in contrast, set female pulses racing, and so, through an enhanced potential for the propagation of genes, may actually represent a successful reproductive strategy.

A cursory peek at the tabloid headlines and gossip-column inches will avail you of the fact that the theory may well hold water. A hell of a lot of it, in fact. But one of the best examples of all, according to Jonason, is James Bond.

"He's clearly disagreeable, very extrovert, and likes trying new things," he points out. "Including killing people. And new women."

Jonason's study saw two hundred college students filling out personality questionnaires specifically designed to assess the presence of Dark Triad attributes. The students were also asked about their sexual relationships, including their attitudes toward casual affairs and one-night stands. Lo and behold, the standout finding was that those who scored higher on the triad tended to have more notches on their rickety, battle-weary bedposts than those who scored lower, suggesting that elements of the three personality styles—narcissism, Machiavellianism, and psychopathy—expedite a dual-process alpha male mating strategy aimed at maximizing reproductive potential:

1. Impregnate as many females as possible.
2. Hit the road before anyone calls you daddy.

And it all seems to have worked out rather well down the years. Otherwise, as Jonason points out, why would such attributes still be knocking around?*

* Though Jonason also found that bad girls get the boys, the relationship between Dark Triad attributes and number of short-term relationships was stronger for men than for women. Of course, the reasons why bad boys might get the girls is another matter. Psychopathy is associated with a lack of neuroticism and anxiety, which may offset fears of rejection and project an air of dominance; narcissism is associated with self-promotion and ostentatious displays of success; and Machiavellianism goes with being socially manipulative. These three traits combined may well,

The Business End of the Psychopathic Spectrum

Curiously, it's not just in terms of reproduction that psychopaths end up on top. The exploits of evolutionary psychologists* such as Peter Jonason add support to the claims of game theory mandarins like Andrew Colman, whom we met in the previous chapter, that there are other areas of life, other fields of endeavor, in which it pays to be a psychopath. A psychopathic strategy doesn't just code for greater success in the bedroom. It also comes in handy in the boardroom.

A 2005 study conducted by a joint team of psychologists and neuroeconomists from Stanford University, Carnegie Mellon University, and the University of Iowa demonstrates this beautifully. The study took the form of a gambling game consisting of twenty rounds. Participants were divided into three groups: normal people, patients with lesions in the emotion areas of the brain (the amygdala, the orbitofrontal cortex, and the right insular or somatosensory cortex), and patients with lesions in brain regions unrelated to emotion. At the start of the game, each participant was handed the sum of $20, and at the beginning of each new round they were asked whether they were prepared to risk $1 on the toss of a coin. While a loss incurred the penalty of $1, a win swelled the coffers by a cool $2.50.

It doesn't take a genius to work out the winning formula. "Logically," says Baba Shiv, professor of marketing at Stanford Graduate School of Business, "the right thing to do is to invest in every round."

But logic, as the political activist Gloria Steinem once remarked, is often in the eye of the logician.

If, as game theory predicts, there are times when it really does pay to keep our foot on the gas—and psychopaths have heavier boots—then,

---

over the short term, give the impression of a cool, confident, and charismatic individual who's fun to be with and who's "going places." Over the long term, however, things often turn out differently.

* Evolutionary psychologists seek to account for human traits and behaviors—such as personality and mating strategies—as functional products of natural selection: as psychological adaptations that evolved to solve recurrent problems in ancestral environments.

according to the dynamics of the game, those participants with the relevant presenting pathology (deficits in emotional processing) should clean up. They should outperform those without, that is, both of the other groups.

This is exactly how the study panned out. As the game unfolded, participants with normal emotional brain centers (whether or not they had dings elsewhere) began declining the opportunity to gamble, opting, instead, for the bewilderingly conservative alternative: holding on to their winnings. In contrast, however, those whose brains were not equipped with the everyday emotional seat belts that most of us keep tightly fastened just kept on rolling, ending the game with a significantly higher profit margin than their opposite numbers.

"This may be the first study," comments George Loewenstein, professor of economics and psychology at Carnegie Mellon, "that documents a situation in which people with brain damage make better financial decisions than normal people."

Antoine Bechara, now professor of psychology and neuroscience at the University of Southern California, goes one better. "Research needs to determine the circumstances in which emotions can be useful or disruptive, [in which they] can be a guide for human behavior," he points out. "The most successful stockbrokers might plausibly be termed 'functional psychopaths'—individuals who on the one hand are either more adept at controlling their emotions or who, on the other, do not experience them to the same degree of intensity as others."

And Baba Shiv agrees. "Many CEOs," he adds rather unnervingly, "and many top lawyers might also share this trait."

A study conducted by the economist Cary Frydman and his colleagues at the California Institute of Technology lends credibility to Shiv's observations. Frydman handed volunteers a sum of $25 and then presented them with a series of tricky financial dilemmas. Within a short, set period, the volunteers had to decide whether to play it safe and accept a sure thing—say, receive $2—or whether to gamble and go for a riskier, but potentially more lucrative, option: a 50-50 chance of gaining $10 or losing $5, for example. Who would clean up, and who would go bust?

Far from it being a matter of random chance, it turned out that a subset of volunteers completely outsmarted the rest, making consistently optimal choices under risk. These individuals were not financial whiz kids. Nor were they economists, mathematicians, or even World Series of Poker champions. Instead, they were carriers of the "warrior gene"—a monoamine oxidase A polymorphism called MAOA-L, previously (if controversially) associated with dangerous, "psychopathic" behavior.

"Contrary to previous discussion in the literature, our results show these behavioral patterns are not necessarily counterproductive," Frydman's team wrote, "since in the case of financial choice these subjects engage in more risky behavior only when it is advantageous to do so."

Frydman elucidated further. "If two gamblers are counting cards and one is making a lot of bets," he observed, "it may look like he's more aggressive or impulsive. But you don't know what cards he's counting—he may just be responding to good opportunities."

Additional support comes from work carried out by Bob Hare and his colleagues in 2010. Hare handed out the PCL-R to more than two hundred top U.S. business executives, and compared the prevalence of psychopathic traits in the corporate world to that found in the general population at large. Not only did the business execs come out ahead, but psychopathy was positively associated with in-house ratings of charisma and presentation style: creativity, good strategic thinking, and excellent communication skills.

Then, of course, there was the survey conducted by Belinda Board and Katarina Fritzon that we discussed in chapter 1. Board and Fritzon pitted company CEOs against the inmates of Broadmoor Hospital, a high security forensic institution in the U.K. (which we'll be getting into, quite literally, in more detail later on), on a psychological profiling test. Once again, when it came to psychopathic attributes the CEOs emerged victorious—which, considering that Broadmoor houses some of Britain's most dangerous criminals, is really going some.

I put it to Hare that in recent years the corporate environment, with its downsizing, restructuring, mergers, and acquisitions, has actually become even more of a hothouse for psychopaths. Just as political turmoil and uncertainty can make for a pretty good petri dish in which to cultivate psychopathy, so, too, I opined, can the high seas of trade and industry. He nodded.

"I've always maintained that if I wasn't studying psychopaths in prison, I'd do so at the stock exchange," he enthused. "Without doubt, there's a greater proportion of psychopathic big hitters in the corporate world than there is in the general population. You'll find them in any organization where your position and status afford you power and control over others, and the chance of material gain."

His coauthor on the corporate psychopathy paper, New York industrial and organizational psychologist Paul Babiak, agrees.

"The psychopath has no difficulty dealing with the consequences of rapid change. In fact, he or she thrives on it," he explains. "Organizational chaos provides both the necessary stimulation for psychopathic thrill seeking and sufficient cover for psychopathic manipulation and abusive behavior."

Ironically, the rule-bending, risk-taking, thrill-seeking individuals who were responsible for tipping the world economy over the edge are precisely the same personalities who will come to the fore in the wreckage. Just like Frank Abagnale, they are the mice who fall into the cream, fight and fight, and churn that cream into butter.

Champagne on Ice

Babiak and Hare's pronouncements—like Board and Fritzon's, demographic and sociological in nature—provide plenty of food for thought. And when placed alongside more empirically derived observations, the fiscal fandangos of neuroeconomists such as Baba Shiv and his coauthors, the coital correlations of Dark Triad hunter Peter

Jonason, and the mathematical machinations of game theorists like Andrew Coleman, for example, they show beyond doubt that there's most definitely a place for the psychopath in society.

This explains, in part, why psychopaths are still around—the inexorable perseverance of their dark, immutable gene streams—and why the evolutionary share price in this niche personality consortium has remained stable and buoyant over time. There are positions in society, jobs and roles to fulfill, which, by their competitive, cutthroat, or chillingly coercive natures, require access to office space in precisely the kind of psychological real estate that psychopaths have the keys to, that they have on offer in their glossy neural portfolios. Given that such roles—predominantly by virtue of their inherent stress and danger—often confer great wealth, status, and prestige on the individuals who assume them, and that, as Peter Jonason showed us, bad boys seem to have a way with certain girls, it's really not surprising that the genes have hung about. Biologically, you might say, they punch above their weight.

Of course, similar charisma and coolness under pressure can also be found among those who take advantage of society—such as the world's top con artists. And, when combined with a genius for deception, this jaw-dropping profile can be devastating. Take Greg Morant. Morant is one of America's most successful and elusive con men—and when it comes to psychopaths, among the top five most charming and the top five most ruthless I've ever had the pleasure of meeting. I caught up with him in the bar of a five-star hotel in New Orleans. It was only after he bought the drinks, a bottle of Cristal champagne for $400, that he handed me back my wallet.

"One of the most important things that a grifter must have in his possession is a good . . . 'vulnerability' radar," Morant illuminated, in a comment reminiscent of the work of psychologist Angela Book. (If you recall from chapter 1, Book found that psychopaths were better than non-psychopaths at discerning the victims of a previous violent assault simply from the way they walked.) "Most folk you come across pay no attention to what they say when they're talking to you. Once out, the words are gone. But a grifter will zone in on everything . . .

Like therapy, you're trying to get inside the person. Figure out who they are from the little things. And it's always the little things. The devil's in the detail . . . You get them to open up. Usually by telling them something about yourself first—a good grifter always has a narrative. And then immediately change the subject. Randomly. Abruptly. It can be anything . . . some thought that just occurred to you out of the blue or whatever . . . anything to interrupt the flow of conversation. Nine times out of ten the person will completely forget what they've just said.

"Then you can get to work—not right away, you need to be patient. But a month or two later. You modify whatever it is, whatever the hell they've told you—you tend to know instantly where the pressure points are—and then tell the story back as if it were your own. Bam! From that point on, you can pretty much take what you want.

"I'll give you an example . . . [One guy is] rich, successful, works like a dog . . . When he's a kid, he comes home from school to find his record collection gone. His pop's a bum and has sold it to stock up his liquor cabinet. He's been collecting these records for years.

"So wait, I think. You're telling me this after, what, three or four hours in a bar? There's something going down. Then I get it. So that's why you work so goddamned hard, I think. It's because of your pappy. You're scared. You're life's been on hold all these years. You're not a CEO. You're that scared little kid. The one who's going to come home from school one day and find your record collection is history.

"Jesus, I think! That's hilarious! So guess what? A couple of weeks later I tell him what happened to me. How I get home from work one night and find my wife in bed with the boss. How *she* files for divorce. And cleans *me* out."

Morant pauses, and pours us some more champagne.

"Total bullshit!" he laughs. "But you know what? I did that guy a favor. Put him out of his misery. What do they say—the best way to overcome your fears is to confront them? Well, someone had to be Daddy."

Morant's words are chilling. Even more so when you hear them firsthand. At close quarters. I distinctly remember our meeting in

New Orleans, and how I felt at the time. Violated, but captivated. Enthralled, but creeped out—much like the clinicians and law enforcement agents that Reid Meloy interviewed, back in chapter 1. I was under precious few illusions as to the kind of man I was dealing with, despite his style and the millionaire yachtsman vibe. Here, in all his glory, was a psychopath. A predatory social chameleon. As the champagne flowed, and the slow southern twilight glinted off his Rolex, he would colonize your brain synapse by synapse without even breaking a sweat. And without your even knowing.

And yet, as a psychologist I saw the simple, ruthless genius in what Morant was saying. His modus operandi adheres to strict scientific principles. Research shows that one of the best ways of getting people to tell you about themselves is to tell them something about yourself. Self-disclosure meets reciprocity. Research also shows that if you want to stop someone from remembering something, the key is to use distraction. And, above all, to use it fast.* And in clinical psychology, there comes a point in virtually every therapeutic intervention where the therapist strikes gold: uncovers a time, a defining moment or incident, that either precipitates the underlying problem or encapsulates it, or both. And this doesn't just apply to dysfunction. Core personality structures, interpersonal styles, personal values—all these things are often best revealed in the small print of people's lives.

"Whenever you interview someone, you're always on the lookout for the seemingly inconsequential," says Stephen Joseph, professor of psychology, health, and social care in the Centre for Trauma,

* Back in the 1950s, the American memory researchers John Brown and Lloyd and Margaret Peterson conducted studies in which participants were given groups of letters to remember and, at the same time or immediately afterward, presented with a numerical distraction. For example, subjects told to remember a three-letter syllable were quickly given a random three-digit number (e.g., 806) and asked to count backward in threes from it. Then, at various intervals, they were asked to recall the letters they'd been given. A control group was shown the letters without the distraction task. Which group did better at recall? Correct: the group that wasn't distracted. In fact, for the group that *was* distracted, memory was totally erased after only 18 seconds. (Brown, 1958; Peterson and Peterson, 1959).

Resilience and Growth at the University of Nottingham. "The flare-up in the office ten years ago with Brian from Accounts. The time when the teacher said you were late and couldn't join in. Or when you did all the work and what's-his-face took the credit. You're looking for needles, not haystacks. The shrapnel of life trapped deep within the brain."

What was that about you doing all the work and someone else taking the credit? Surely not.

The Truth About Lying

Con artist and secret agent are two sides of the same coin, if the views of one of the U.K.'s senior homeland security figures that I spoke to are anything to go by. Both, she pointed out, rely on the ability to pass oneself off as something one is not, the facility to think on one's feet, and the capacity to navigate webs of deception with alacrity.

I'd be surprised if Eyal Aharoni would disagree. In 2011, Aharoni, a psychology postdoc at the University of New Mexico, asked a question that, hard though it is to believe, no one had asked before. If, under certain conditions, psychopathy really is beneficial, then does it make you a better criminal? To find out, he sent out a survey to more than three hundred inmates in a bunch of medium-security prisons across the state. Computing a "criminal competence" score for each inmate by comparing the number of crimes committed with their total number of non-convictions (e.g., 7 non-convictions out of a total of 10 crimes = 70 percent success rate), Aharoni uncovered something interesting: psychopathy does indeed predict criminal success. That said, there's a limit. A very high dose of psychopathy (all the dials turned up to max) is as bad as a very low one. Instead, it's moderate levels that code for greater "accomplishment."

Precisely how psychopathy makes one a better criminal is open to debate. On the one hand, psychopaths are masters at keeping their cool under pressure, which may well give them an edge in a getaway car or an interview room. On the other hand, they're also ruthless, and might intimidate witnesses into not coming forward with evidence. But equally plausible—and equally apposite to spies and grifters alike—is that as well as being ruthless and fearless, psychopaths are in possession of another, more refined personality talent. Exactly like the world's top poker players, they might also be better at controlling their emotions than others when the stakes are high and backs are against the wall, which would give them an edge not just outside the courtroom, when planning and effecting their nefarious schemes and activities. But inside it as well.

Up until 2011, the evidence for this was largely circumstantial. Helinä Häkkänen-Nyholm, a psychologist at the University of Helsinki, had observed, in conjunction with Bob Hare, that psychopathic offenders appeared more convincing than non-psychopathic offenders when it came to expressing remorse—which is odd, to say the least, because it's something they're unable to feel. But a quick look at the context of such observations—before the court, just prior to sentencing; before the court, to appeal a sentence; and before psychologists and prison governors at parole board hearings—aroused the suspicions of psychologist Stephen Porter. The issue was one of "affective authenticity." Remorse aside, Porter wondered, were psychopaths just better at faking it?

Porter and his colleagues devised an ingenious experiment. Volunteers were presented with a series of images that were designed to evoke various emotions, and then responded to each with either a genuine or a deceptive expression. But there was a catch. As the participants viewed the emotionally charged pictures, Porter videotaped them at a speed of thirty frames per second and then examined the tapes frame by frame. This, in the "deception" condition, allowed him to screen for the presence of physiognomical lightning strikes called "microexpressions": fleeting manifestations of true, unadulterated emotion—invisible, in real time, to most people's naked eye—that flash, imperceptibly, through the shutters of conscious concealment (see figure 4.1).

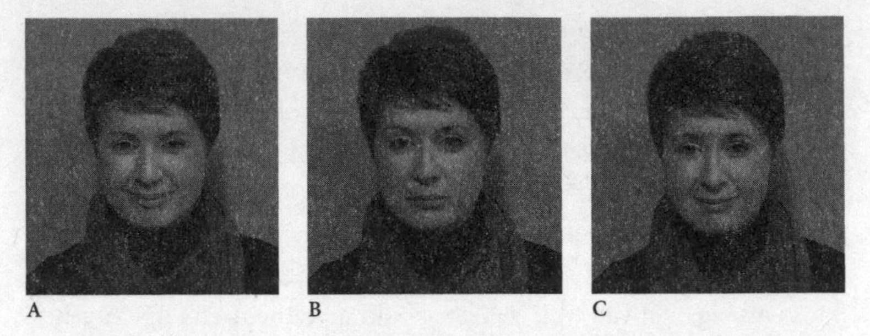

Figure 4.1. Picture A shows a genuine smile, while picture C shows a false smile with leakage of sadness (lowering of the eyebrows, eyelids, and corners of the mouth). Picture B shows a neutral expression. Even minuscule—and fleeting—changes such as this are able to alter the entire face.

Porter wanted to know if participants exhibiting higher levels of psychopathy would be more adept at disguising the true nature of their feelings than their lower-scoring counterparts. The answer, unequivocally, was yes. The presence (or absence) of psychopathic traits significantly predicted the degree of inconsistent emotion observed in the deception condition. Psychopaths were far more convincing at feigning sadness when presented with a happy image, or happiness when looking at a sad image, than were non-psychopaths.* Not only that, but they were as good as volunteers who scored high on emotional intelligence. If you can fake sincerity, as someone once said . . . well, you really have got it made, it would seem.

Cognitive neuroscientist Ahmed Karim has taken things one stage further—and with the aid of some electromagnetic magic can significantly improve the career prospects of both con artists *and* secret agents. Karim and his team at the University of Tübingen, in Germany, can make you a better liar. In an experiment in which volunteers role-played stealing money from an office and were then interrogated by a researcher acting as a police detective (as an incentive to deceive the detective, the would-be "thieves" were allowed to keep the money if successful!), Karim discovered that the application

* Intriguingly, one of Porter's students, Sabrina Demetrioff, has also found the opposite: that psychopaths are better at decoding microexpressions in others.

of a technique known as transcranial magnetic stimulation (TMS)* to the part of the brain implicated in moral decision making, the anterior prefrontal cortex, elevated participants' lying quotient. It gave them a higher Lie Q.

Precisely why this should be the case is not immediately obvious, and researchers are considering their options. But one possibility is that TMS-induced inhibition of the anterior prefrontal cortex implements the restriction of a neural no-fly zone over conscience, sparing the liar the distractions of moral conflict. Such a hypothesis is consistent with research on psychopaths. We know from previous studies, for instance, that psychopaths have reduced gray matter in the anterior prefrontal cortex—and recent analysis using diffusion tensor imaging (DTI),† conducted by Michael Craig and his coworkers at the Institute of Psychiatry in London, has also revealed reduced integrity of the uncinate fasciculus: the axonal tract (a kind of neural aqueduct) connecting the prefrontal cortex and amygdala.

Psychopaths, in other words, not only have a natural talent for duplicity, but also feel the "moral pinch" considerably less than the rest of us. Not always a bad thing when the chips are down and decisions must be made under fire.

*TMS is a noninvasive method of temporarily stimulating the brain in order to disrupt cortical processing, and thereby to investigate the effects of either exciting or inhibiting selected neural pathways.

†DTI tracks the movement of water molecules in the brain. In most brain tissue, as in most other kinds of tissue, the diffusion of water molecules is multidirectional. In the tracts of white matter, however—the bundles of fibers that conduct electrical impulses between different areas of the brain—water molecules tend to diffuse directionally along the length of the axons, the long, slender filaments that project outward from the base of each neuron, conducting electrical impulses away from the cell body toward synapses with receiving cells. Axons have an insulating and "waterproofing" coat of white, fatty myelin—it's what makes the white matter white—which can vary in thickness. Thus, by analyzing the rate and direction of water diffusion, researchers can create "virtual" pictures of axons, make inferences as to the thickness of these white myelin coats, and assess their structural integrity.

Cool of the Moment

Of course, it isn't just liars who benefit from a dearth of morality. The ethically challenged may be found in all walks of life—not just in casinos and courtrooms. Take, for instance, the following exchange from the 1962 film *The War Lover*:

LIEUTENANT LYNCH: Now, what about Rickson? We never know what stunt he'll pull next. Can we afford to have that sort of pilot? Can we afford not to have him? What's your opinion, doc?

CAPTAIN WOODMAN: Rickson's an example of the fine line that separates the hero from the psychopath.

LIEUTENANT LYNCH: Which side of the line do you place Rickson?

CAPTAIN WOODMAN: Time will tell. I suppose we're running a risk . . . but then that's the nature of war.

The War Lover, set in World War II, features a character called Buzz Rickson, an arrogant, fearless B-17 pilot, whose genius at aerial combat provides the perfect outlet for his ruthless, amoral dark side. When a bombing mission is aborted due to adverse weather conditions, Rickson, much feted by his crew for his daredevil flying skills, disobeys the order to turn around, diving under the cloud cover to drop his deadly cargo. Another of the bombers fails to return to base. Rickson's elemental, predatory instincts revel in the theater of war. Assigned by his commanding officer to a routine sortie dropping propaganda leaflets, he buzzes the airfield in protest, setting the scene for the above dialogue between his navigator and the flight surgeon.

It's a fine line, as Captain Woodman says, between hero and psychopath. And often it depends who's drawing it.

Characters like Rickson don't exist just in the movies. Of a number of Special Forces soldiers I've tested so far, all of them have scored high on the PPI—which is no real surprise given some of the things they get into. As one of them, with characteristic understatement, put it: "The lads who took out Bin Laden weren't on some paintballing weekend . . ."

Such coolness and focus is illustrated in a study conducted by the psychologist and neuroscientist Adrian Raine and his colleagues at the University of Southern California in Los Angeles. Raine compared the performance of psychopaths and non-psychopaths on a simple learning task, and found that when mistakes were punished by a painful electric shock, psychopaths were slower to pick up the rule than non-psychopaths. But that was just the half of it. When success was rewarded by financial gain, as well as by avoidance of shock, the roles reversed. This time it was the psychopaths who were quicker on the uptake.

The evidence is pretty clear. If the psychopath can "make" out of a situation, if there's any kind of reward on offer, they go for it, irrespective of risk or possible negative consequences. Not only do they keep their composure in the presence of threat or adversity, they become, in the shadow of such presentiment, laser-like in their ability to "do whatever it takes."

Researchers at Vanderbilt University have delved a little deeper, and have looked at how the unblinking, predatory focus commonly displayed by psychopaths might actually be mirrored in their brains. What they've discovered sheds a completely different light on how it might feel to be a psychopath, and, as such, opens up a whole new perspective on precisely what makes them tick. In the first part of the study, volunteers were divided into two groups: those exhibiting high levels of psychopathic traits and those on the low side. The researchers then gave both groups a dose of speed (otherwise known as amphetamine) and, using positron emission tomography (PET),* scrutinized their brains to see what might unfold.

"Our hypothesis was that [some] psychopathic traits [impulsivity, heightened attraction to rewards, and risk taking] are . . . linked to dysfunction in dopamine reward circuitry," elucidates Joshua

* PET allows researchers to obtain images of discrete neurochemical activity in assorted areas of the brain as subjects engage in different activities, thoughts, or emotions. This is achieved by injecting a harmless and short-lived radioactive dye into the bloodstream of a volunteer, and then tracking the destination of the dye by mapping the radiation emission patterns in the form of gamma rays.

Buckholtz, the lead author of the study, "[and] that because of these exaggerated dopamine responses, once they focus on the chance to get a reward, psychopaths are unable to alter their attention until they get what they're after."

He wasn't far off the mark. Consistent with such a hypothesis, the volunteers displaying high levels of psychopathic traits released almost four times as much dopamine in response to the stimulant as did their non-psychopathic counterparts. But that wasn't all. A similar pattern of brain activity was observed in the second part of the experiment, when instead of being given speed, the participants were told that, on completion of a simple task, they'd receive a monetary reward. (Note to the researchers: if you need any more volunteers, call me!) Sure enough, fMRI revealed that those individuals with elevated psychopathic traits exhibited significantly more activity in their nucleus accumbens, the dopamine reward area of the brain, than those scoring low on psychopathy.

"There has been a long tradition of research on psychopathy that has focused on the lack of sensitivity to punishment and a lack of fear," comments David Zald, associate professor of psychology and psychiatry, and coauthor of the study. "But those traits are not particularly good predictors of violence, or criminal behavior . . . These individuals appear to have such a strong draw to reward—to the carrot—that it overwhelms the sense of risk or concern about the stick . . . It's not just that they don't appreciate the potential threat, but that the anticipation or motivation for reward overwhelms those concerns."

Corroborating evidence comes from forensic linguistics. The way a murderer talks about his crime depends, it turns out, on what type of murderer he is. Jeff Hancock, a professor of computing and information science at Cornell, and his colleagues at the University of British Columbia compared the accounts of fourteen psychopathic and thirty-eight non-psychopathic male murderers, and uncovered notable differences: not just in relation to emotional pixilation (the psychopaths used twice as many words relating to physical needs, such as food, sex, or money, as the non-psychopaths, who placed more of an emphasis on social needs, such as family, religion, and spirituality), but also in relation to personal justification.

Computer analysis of taped transcripts revealed that the psychopathic killers used more conjunctions like "because," "since," or "so that" in their testimonies, implying that the crime somehow "had to be done" in order to attain a particular goal. Curiously, they also tended to include details of what they'd had to eat on the day of the murder—the spectral machinations of the hand of primeval predation?

Be that as it may, the conclusion is little in doubt. The psychopath seeks reward at any cost, flouting consequence and elbowing risk aside. Which, of course, might go some way toward explaining why Belinda Board and Katarina Fritzon found a greater preponderance of psychopathic traits among a sample of CEOs than they did among the inmates of a secure forensic unit. Money, power, status, and control—each the preserve of the typical company director, and each a sought-after commodity in and of itself—together constitute an irresistible draw for the business-oriented psychopath as he or she ventures ever further up the rungs of the corporate ladder. Recall, from earlier, that stark, prophetic caveat of Bob Hare's: "You'll find them [psychopaths] in any organization where your position and status afford you power and control over others, and the chance of material gain."

Sometimes they do a good job. But sometimes, inevitably, they don't. And if the reward ethic gets out of hand, the boom, rather predictably, can quickly turn to bust. Arrogant and fearless Buzz Ricksons may be found all over the place, in pretty much any field you can think of. Including, oddly enough, banking.

And Rickson, in case you were wondering, ended up dead: crashing, in an inglorious ball of flames, into the white cliffs of Dover.

Hot Reading

The psychopath's fearlessness and focus has traditionally been attributed to deficits in emotional processing, more specifically to amygdala dysfunction. Until recently, this has led researchers to believe that in addition to not "doing" fear, they don't "do" empathy, either. But a 2008 study by Shirley Fecteau and her colleagues at the Beth Israel Deaconess Medical Center in Boston, has thrown a completely different light on the matter, suggesting that psychopaths not only have the capacity to recognize emotions—they are, in fact, actually better at it than we are.

Fecteau and her coworkers used TMS to stimulate the somatosensory cortex (the part of the brain that processes and regulates physical sensations) in the brains of volunteers scoring high on the PPI. Previous research has shown that observing something painful happening to someone else results in a temporary slowdown in neural excitation in response to TMS, in the area of the somatosensory cortex corresponding to the region afflicted by the pain: the work of highly specialized, and aptly named, brain structures called mirror neurons. If psychopaths lack the ability to empathize, Fecteau surmised, then such attenuation in neural response should be reduced in those individuals scoring high on the PPI, compared to those with low to average scores—in exactly the same way that psychopaths might well, in comparison with most normal members of the population, display reduced yawn contagion.*

The researchers, however, were in for quite a surprise. Much to their amazement, Fecteau and her team actually turned up the op-

* Yawn contagion is indicative of a deep bodily connection between humans, between other animals, and even, in some cases, between humans *and* other animals! Dogs "catch" yawns from their owners, and chimpanzees also catch yawns from their handlers. The general consensus centers around two possibilities. Either individuals who have problems with empathy show reduced yawn contagion because they don't pay attention to the yawns of others. Or, alternatively, they're simply not affected by them. A colleague, Nick Cooper, and I are currently in the process of testing the yawn reflex in psychopaths in an ongoing study in Sweden.

posite of what they were expecting. High PPI scorers—specifically, those who scored high on the "Coldheartedness" subscale of the questionnaire, the subscale that most directly taps into empathy—in fact showed greater attenuation of the TMS response than low scorers, suggesting that psychopaths, rather than having an impairment in recognizing the emotions of others, indeed have a talent for it. And that the problem lies not in emotional recognition per se, but in the dissociation between its sensory and affective components: in the disconnect between knowing what an emotion is and feeling what it's like.

Psychologist Abigail Baird has discovered something similar. In an emotion recognition task using fMRI, she found that while volunteers scoring high on the PPI showed reduced amygdala activity compared to low scorers when matching faces with similar emotional expressions (consonant with a deficit in emotional processing), they also displayed increased activity in both the visual and dorsolateral prefrontal cortices—indicative, as Baird and her team point out, of "high-scoring participants relying on regions associated with perception and cognition to do the emotion recognition task."

One psychopath I spoke to put it like this. "Even the color-blind," he said, "know when to stop at a traffic light. You'd be surprised. I've got hidden shallows."

Or, as Homer Simpson reminded us earlier, not caring and not understanding are two different things entirely.

Of course, psychopaths' enhanced ability to recognize emotion in others might go some way toward explaining their superior persuasion and manipulation skills—as, needless to say, does their enhanced ability to fake emotion, a phenomenon we touched upon earlier in the chapter. But the capacity to decouple "cold" sensory empathy from "hot" emotional empathy has other advantages, too—most notably in arenas where a degree of affective detachment must be preserved between practitioner and practice. Like the medical profession, for instance.

Here's one of the U.K.'s top neurosurgeons on how he feels prior to going into the theater: "Do I get nervous before a big operation? No,

I wouldn't say so. But I guess it's like any performance. You have to get yourself psyched up. And you need to remain concentrated and focused on the job at hand, not get distracted. You have to get it right.

"You know, you mentioned Special Forces a few moments ago. And actually, the mentality of a surgeon and, say, that of an elite soldier about to storm a building or an airliner are possibly quite similar. In both cases, the job is referred to as an 'operation.' In both cases, you 'tool up' and don a mask. And in both cases, all the years of practice and training can never fully prepare you for that element of uncertainty as you make the first incision; that exhilarating split second of 'explosive entry,' when you fold back the skin and suddenly realize . . . you're IN.

"What's the difference between a millimeter's margin of error when it comes to taking a head shot and a millimeter's margin of error when it comes to navigating your way between two crucial blood vessels? In both of those cases you hold life and death in your hands, must make a death-or-glory decision. In surgery, quite literally, on a knife-edge."

This guy scored well above average on the PPI. And if that surprises you coming from one of the world's top neurosurgeons, then think again. Yawei Cheng, at the National Yang-Ming University in Taiwan, and her coworkers took a group of medical doctors with at least two years' experience in acupuncture and a group of nonmedical professionals, and, using fMRI, peered into their brains to see what happened as they viewed needles being inserted into mouths, hands, and feet. What they observed was rather interesting. When the control volunteers watched the videos of the needles being inserted, those areas of their somatosensory cortices corresponding to the relevant body regions lit up like Christmas trees, as did other brain areas such as the periaqueductal gray (the coordinator of the panic response) and the anterior cingulate cortex (which codes for error, anomaly, and pain processing).

In contrast, there was barely a flicker of pain-related activity in the brains of the experts. Instead, they exhibited increased activation of the medial and superior prefrontal cortices, as well as of the temporoparietal junction: brain regions involved in emotion regulation and theory of mind.* Moreover, the experts rated the acupuncture displays as significantly less unpleasant than the controls did—reminiscent of numerous laboratory findings showing attenuated physiological responses (e.g., heart rate, galvanic skin response [GSR], and cortisol levels) in psychopaths on presentation of fearful, disgusting, or erotic stimuli—and in the face of arduous social stress tests, such as the Trier.† What the expert acquires through experience, psychopaths have from the start.

Psychopath Lite?

Not long after I came across Yawei Cheng's study, I hopped on a plane to Washington, DC, and went to the National Institute of Mental Health to see James Blair. Blair is one of the world's leading experts on psychopaths, and, like Joe Newman, has pretty much seen it all. "Does it pay to be a psychopath?" I asked him. "Okay, not all the time perhaps. But sometimes—when the situation demands it?"

Blair was cautious. It's a dangerous road to go down. "It's true that if bad things are happening the individual with psychopathy might be less worried about it," he told me. "However, it's not so clear that their decision making in such situations would be particularly good, though. Moreover, by not analyzing levels of threat appropriately, they might walk into danger, rather than away from it."

In other words, if we could somehow defrost the reasoning a bit, take some of the chill out of the logic, then yes, psychopathic traits may well confer advantages. Otherwise, forget it.

But wait a minute, I thought. Isn't this precisely what we find in

* Broadly speaking, theory of mind refers to the ability to see, in both a cognitive and emotional sense, where others are "coming from."

† The Trier Social Stress Test typically involves volunteers being given only a brief amount of time to prepare a mock job talk, during which they are told that they will undergo various kinds of professional scrutiny, such as voice frequency analysis, assessment of nonverbal communication skill, etc.

the heroes of this world? No one would accuse *them* of poor decision making. And what about the performance of Bechara, Shiv, and Loewenstein's "functional psychopaths"? And Frydman's hotshot hustlers? (Okay, carrying the MAOA polymorphism that codes for risk and aggression doesn't automatically qualify you as a psychopath, but the link is certainly there.) The way things turned out, their decision making, under the circumstances, would more likely than not have been better than yours or mine. So maybe that was it. Maybe the equation just needed a bit of loss adjusting:

Functional Psychopath = Psychopath – Poor Decision Making

By way of a second opinion, I caught up with psychopath hunter Kent Kiehl. Kiehl is associate professor of psychology and neuroscience at the University of New Mexico, and director of Mobile Imaging Core and Clinical Cognitive Neuroscience at the Mind Research Network in Albuquerque. As his job titles suggest, he had a lot going on when I met him. In actual fact, Kiehl was on a road trip when we hooked up—and still is. Not your average kind of road trip, but one involving an eighteen-wheeler: a truck so large that every time he parks it I'm amazed he doesn't need planning permission. He certainly needs scanning permission—because inside it is a custom-built fMRI machine, worth $2 million. And Kiehl is carting it around New Mexico, around a number of the state penitentiaries, in an attempt to unravel the neural basis of psychopathy. I asked him the same question I asked James Blair. Does it pay, at times, to be a psychopath? Kiehl, like Blair, was circumspect.

"It certainly makes sense that psychopathic traits are normally distributed across the general population," he told me. "But the difference with those at the high end of the spectrum is that they can't switch off [the fearlessness] in situations where it might be appropriate. A CEO might be non-risk-averse in certain areas of business, but, on the other hand, probably wouldn't want to walk around a rough neighborhood at night. A psychopath isn't able to make that distinction. With a psychopath, it's all or nothing." Which adds a third factor to our equation:

$$\frac{\textbf{Functional Psychopath} = (\textbf{Psychopath} - \textbf{Poor Decision Making})}{\textbf{Context}}$$

Which means, minus the algebra, that functional psychopathy is context dependent. That, in the language of personality theory, it is "state" as opposed to "trait." And that, in the right set of circumstances, it can enhance rather than encumber the speed and quality of decision making.

Back in the 1980s, the sociologist John Ray reached a similar conclusion. Ray postulated an inverted-U-shaped function as the model most befitting the relationship between psychopathy and life success (see figure 4.2). In his own words:

"Both extremely high and extremely low levels of psychopathy may be maladaptive, with intermediate levels being most adaptive. The basis for saying that high levels of psychopathy are maladaptive is, of course, the trouble into which clinical psychopaths often get themselves. The basis for saying that low levels of psychopathy may also be maladaptive stems from the common observation of the role of anxiety in psychopathy: psychopaths do not seem to show any anxiety. The debilitating function of high levels of anxiety hardly needs to be stressed. In a normal, noninstitutionalized population, therefore, their relative immunity from anxiety may give psychopaths an advantage."

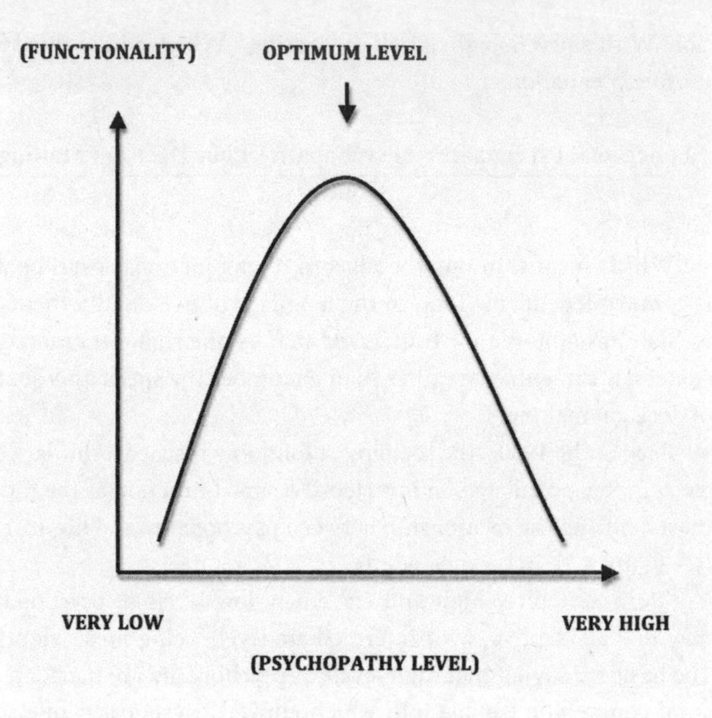

Figure 4.2. The relationship between psychopathy and functionality (from Ray and Ray, 1982)

Ironically, this is precisely what Eyal Aharoni found within the criminal fraternity. It was neither high nor low levels of psychopathy that coded for criminal success—it was moderate levels: something that hasn't escaped the attentions of Bob Hare and Paul Babiak as they continue their research into the area of corporate psychopathy. Hare and Babiak have developed an instrument called the Business Scan (B-Scan for short): a self-report questionnaire consisting of four subscales (personal style; emotional style; organizational effectiveness; and social responsibility) specifically calibrated to assess the presence of psychopathic traits, not within forensic populations (like the PCL-R), or within the general population as a whole (like the PPI), but exclusively within corporate settings (see figure 4.3).

Within such settings, core psychopathic traits can sometimes morph into the star qualities characteristic of an influential leader, and in order to assess the presence of such traits—with due sensitivity to context—it becomes imperative to ask the right questions, employing, of course, the right phraseology and language. The aim of the B-Scan is to do just that, by depicting items within a corporate framework and by couching them in everyday business terminology (e.g., "It's okay to lie in order to clinch the deal"—agree/disagree on a scale of 1 to 4). At present, we're in the process of rolling it out to an independent sample of lawyers, traders, and Special Forces soldiers in the U.K., to see precisely what they're made of: a kind of psychological biopsy of various high-octane professions.

LEADERSHIP TRAIT	PSYCHOPATHIC TRAIT
Charismatic	Superficial charm
Self-confidence	Grandiosity
Ability to influence	Manipulation
Persuasive	Con artistry
Visionary thinking	Fabrication of intricate stories
Ability to take risks	Impulsivity
Action oriented	Thrill seeking
Ability to make hard decisions	Emotional poverty

Figure 4.3—The B-Scan: leadership traits and their psychopathic equivalents.

In a café in upstate New York, just down the road from Babiak's leadership and management consultancy company, I recount a conversation I once had with a top British Queen's Counsel, at his chambers in central London.

"In the courtroom, I have literally rubbed people out," the guy told me, "crucified them in the witness box. I have absolutely no problem at all reducing an alleged rape victim to tears on the stand. You know why? Because that's my job. That's what my client pays me to do. At the end of the day, I can hang up my wig and gown, go out to a restaurant with my wife, and not give a damn—even though I know that what happened earlier might possibly have ruined her life.

"On the other hand, however, if my wife buys a dress from a department store, say, and has lost the receipt and asks me to take it back . . . that's a different story altogether. I hate doing stuff like that. I'm hopeless. A real wuss . . ."

Babiak nods. He knows exactly what I'm getting at. It's exactly the same thing that the B-Scan's designed to get at.

We sip our lattes and stare into the Hudson River. Above the ice-gray water, vast continents of cloud press slowly and remorselessly across a low, tectonic sky.

"What do you reckon?" I ask him. "Do you think we'll find an optimum score on the B-Scan? A golden number that correlates with peak performance?"

He shrugs. "We might," he says. "But my guess is, it will probably be more like a range. And it may vary slightly depending on profession."

I agree. I can't help thinking of Johnny, and where *he* might fall on the scale. James Bond was licensed to kill. But he didn't kill indiscriminately. He killed when he had to. And didn't bat an eye.

Mad, Bad . . . or Supersane?

To wrap things up, I run my theory of optimal, functional psychopathy past a friend of mine. Tom is a member of the British Special Forces, and has worked undercover in some of the hottest, most remote, and most dangerous places on earth. He loves every minute of it. I tell him about the gambling games, the emotional recognition tasks, Ahmed Karim's transcranial magnetic lie-enhancer, and the acupuncturists. Then I tell him about what James Blair, Kent Kiehl, Bob Hare, Paul Babiak, and Peter Jonason have all said.

"What, exactly, are you getting at?" he asks me when I finally put it to him that donning night vision goggles and knife-fighting the Taliban in deep, dark cave complexes in the mountains of northern

Afghanistan might not be every soldier's cup of tea. "That I'm mad? That I'm some kind of loon who dares to rush in where angels fear to tread? And gets off on it? Gets paid for it?"

Once I'm out of the headlock, Tom tells me a story. One night, a few years ago, he's heading back to his apartment after watching *Saw*. Suddenly, out of a doorway, a guy with a blade appears. His girlfriend is terrified and starts hyperventilating. But Tom, ahem, calmly disarms the guy and sends him packing.

"Funny thing was," Tom says, "I actually thought the movie was quite scary. But then, you know, when I suddenly found myself in a real-life situation, I just, kind of, switched on. There was nothing to it. No nerves. No drama. Just . . . nothing."

The neurosurgeon we heard from earlier agrees. Bach's *St. Matthew Passion* regularly moves him to tears. And when it comes to soccer and the team he's supported since he was a kid . . . sometimes he just can't look.

"Psychopath?" he says. "I'm not sure about that. I'm not sure what my patients would think about it either! But it's a good word. And yes, when you're scrubbing up before a difficult operation, it's true: a chill does go through the veins. The only way I can describe it is to compare it to intoxication. Only it's an intoxication that sharpens, rather than dulls, the senses; an altered state of consciousness that feeds on precision and clarity, rather than fuzziness and incoherence . . . Perhaps 'supersane' would be a better way of describing it. Less sinister. More, I don't know, spiritual . . ."

He laughs. "Then again, maybe that sounds even crazier."

FIVE

MAKE ME A PSYCHOPATH

> The great epochs of our lives are the occasions when we gain the courage to rebaptize our evil qualities as our best qualities.
>
> —FRIEDRICH NIETZSCHE

The Times They Are A-Changin'

When you've been at the top of your game for as long as Bob Hare has, you're entitled to be a little bit choosy about who you hang out with at conferences. So when it came around to the Society for the Scientific Study of Psychopathy's biennial bash in Montreal in 2011, and e-mailing the distinguished professor to fix up a meeting, I felt it best to keep things formal. Were a time window to present itself during the course of the proceedings, how about a coffee, I intimated?

The response was instantaneous. "I prefer fine Scotch to coffee," Hare shot back. "You'll find me in the hotel bar. I'll buy."

He was right. On all three counts.

I decide to start off cautiously. "So what the hell do *you* score on the PCL-R, then, Bob?" I ask, over a twenty-year-old single malt.

He laughs.

"Oh, very low," he says. "Around 1 or 2. My students tell me that I really should work on it a bit more. But I did do something 'psychopathic' not so long ago. I splashed out on a brand-new sports car. A BMW."

"Great!" I say. "Maybe your students have more of an effect on you than you realize."

My second question is more serious: "When you look around you at modern-day society, do you think, in general, that we're becoming more psychopathic?"

This time, he takes a bit longer to answer. "I think, in general, yes, society *is* becoming more psychopathic," he says. "I mean, there's stuff going on nowadays that we wouldn't have seen twenty, even ten, years ago. Kids are becoming anesthetized to normal sexual behavior by early exposure to pornography on the Internet. Rent-a-friend sites are getting more popular on the Web, because folks are either too busy or too techy to make real ones. And I read a report the other day that linked a significant rise in the number of all-female gangs to the increasingly violent nature of modern video game culture. In fact, I think if you're looking for evidence that society is becoming more psychopathic, the recent hike in female criminality is particularly revealing. And don't even get me started on Wall Street!"

Hare's position makes a great deal of sense to those who take even a passing interest in what they read in the newspapers, see on TV, or stumble upon online. In Japan in 2011, a seventeen-year-old boy parted with one of his own kidneys so he could go out and buy an iPad. In China, following an incident in which a two-year-old baby was left stranded in the middle of a marketplace and run over, not once but twice, as passersby went casually about their business, an appalled electorate has petitioned the government to pass a "Good Samaritan" law to prevent such a thing from ever happening again.

On the other hand, however, bad things have always happened in society. And no doubt always will. Harvard psychologist Steven Pinker has recently flagged this in his book *The Better Angels of Our Nature*. In fact, he goes one step further. Far from being on the increase, Pinker argues, violence is actually in decline. The reason that vicious slayings and other horrific crimes make the front pages of our papers isn't because they're commonplace. But rather, the complete reverse.

Take homicide, for instance. Trawling through the court records of a number of European countries, scholars have computed that rates have fallen dramatically down the years. In fourteenth-century Oxford, for example, it seemed, relative to today, that everyone was at it, the rate back then being 110 murders per 100,000 people per year, compared to just 1 murder per 100,000 people in mid-twentieth-century London. Similar patterns have also been documented elsewhere—in Italy, Germany, Switzerland, the Netherlands, and Scandinavia.

The same goes for war. Pinker calculates that even in the conflict-ravaged twentieth century, around 40 million people died on the battlefield out of the approximately 6 billion who lived—which equates to a figure of just 0.7 percent. Incorporate, into that estimate, the war-related demise of those who died from disease, famine, and genocide and the death toll rises to 180 million. That sounds like a lot, but statistically speaking, it's still pretty insignificant, weighing in, give or take, at a modest 3 percent.

Contrast this with the corresponding figure for prehistoric societies—a whopping 15 percent—and you begin to get the picture. The beat-up Neanderthal skull that Christoph Zollikofer and his colleagues dug up in southwest France, if you recall, is just the tip of the iceberg.

The questions, of course, that immediately spring to mind when one is confronted by such figures are twofold. Firstly, do they fit with the idea that society is becoming more, and not less, psychopathic? Secondly, if less, what has happened in the intervening years to still, so dramatically, our murderous, violent impulses?

Taking the latter question first, the obvious answer, or at least the one that probably tumbles most readily out of the majority of people's explanation closets, is law. In 1651, in *Leviathan*, it was Thomas Hobbes who first advanced the contention that without top-down state controls we'd turn, rather effortlessly, into a bunch of brutish savages. There's more than a grain of truth in such a notion. But Pinker argues from a more bottom-up perspective, and while certainly not denying the importance of legal restraints, also insinuates a gradual process of cultural and psychological maturation:

"Beginning in the eleventh or twelfth [century], and maturing in the seventeenth and eighteenth, Europeans increasingly inhibited their impulses, anticipated the long-term consequences of their actions, and took other people's thoughts and feelings into consideration," he points out. "A culture of honor—the readiness to take revenge—gave way to a culture of dignity—the readiness to control one's emotions. These ideals originated in explicit instructions that cultural arbiters gave to aristocrats and noblemen, allowing them to differentiate themselves from the villains and boors. But they were then absorbed into the socialization of younger and younger children, until they became second nature. The standards also trickled down from the upper classes to the bourgeoisie that strove to emulate them, and from them to the lower classes, eventually becoming a part of the culture as a whole."

From both a historic and a sociological perspective, this makes perfect sense. Yet concealed within Pinker's observations are a couple of critical precepts with more immediate implications: subtle, sociobiological clues that, if examined in greater detail, may help square the circle of an interesting cultural paradox and go some way to answering the first of our two questions: the perception of a society on the one hand becoming increasingly less violent, and a society on the other hand becoming apparently more psychopathic.

Consider, for example, in Pinker's elegant exposition, the importance of the "cultural arbiter" as a conduit of ideological change. Traditionally, in days gone by, such arbiters would typically have been clergymen. Or philosophers. Or poets. Or even, in some cases, monarchs. Today, however, with society becoming ever more secularized, and with the exponential expansion of an infinitely virtual universe, they're a different breed entirely: pop stars and actors and media and video game moguls, who, rather than disseminating the dictates of dignity, now offer them up on an altar of creative psychopathy.

You've just got to turn on the television. On NBC's *Fear Factor* we see grossed-out contestants devouring worms and insects. On *The Apprentice* we hear the casual admonition: "You're fired." Simon Cowell wasn't exactly known for his eggshell-walking abilities, was he? And I shudder to think what's happening in Ann Robinson's pants as she fixes a losing player in her prurient, surgically enhanced gaze and announces, like some demented dominatrix diva: "You are the weakest link. Goodbye."

But the cultural transmission of normative strains of behavior constitutes just one side of Pinker's sociobiological equation. Their absorption into society as conventional codes of conduct, until such time as they "become second nature," is another matter entirely. Take the finance industry, for instance . . . Greed and corruption have always encroached upon the fringes of big business—from Civil War profiteers in the U.S. to the insider-trading scandals that simmered beneath the surface of capitalist Thatcherite Britain back in the 1980s. Yet the new millennium has seemingly ushered in a wave of corporate criminality like no other. Investment scams, conflicts of interest, lapses of judgment, and those evergreen entrepreneurial party tricks of good old fraud and embezzlement are now utterly unprecedented. Both in scope and in fiscal magnitude.

Corporate-governance analysts cite a confluence of reasons for today's sullied business climate. Avarice, of course—the backbone of Gordon Gekkoism—is one of them. But also in the mix is so-called guerrilla accounting. As Wall Street and the London Stock Exchange expected continued gains and the speed and complexity of business increased exponentially, rule-bending and obfuscation suddenly became de rigueur.

"With infinitely more complex securities, accounting practices, and business transactions," observes Seth Taube, a senior commercial litigation lawyer, "it's much easier to hide fraud."

Clive R. Boddy, a former professor at the Nottingham Business School, comes right out and says it—and in a recent issue of the *Journal of Business Ethics*, contends that it's psychopaths, pure and simple, that are at the root of all the trouble. Psychopaths, Boddy explicates, in language somewhat reminiscent of that used by Bob Hare and Paul Babiak in the previous chapter, take advantage of the "relatively chaotic nature of the modern corporation," including "rapid change, constant renewal" and the high turnover of "key personnel"—circumstances that not only permit them to wend their way, through a combination of "extroverted personal charisma and charm," to the corner offices of major financial institutions, but that also render "their behavior invisible," and, even worse, make them "appear normal and even to be ideal leaders."

Of course, once in situ, such corporate Attilas are then, according to Boddy's analysis, "able to influence the moral climate of the whole organization" and wield "considerable power." He closes with a damning indictment. It is psychopaths, he concludes, who are to blame for the global financial crisis, because their "single-minded pursuit of their own self-enrichment and self-aggrandizement to the exclusion of all other considerations has led to an abandonment of the old-fashioned concept of noblesse oblige, equality, fairness, or of any real notion of corporate social responsibility." There's no denying he might well be onto something.

On the other hand, however, there's society in general, proclaims Charles Elson, head of the Weinberg Center for Corporate Governance at the University of Delaware—who proposes that rather than laying the blame solely at the door of the corporate fat cats, it should, instead, also be pinned on a culture of moral malfeasance, in which truth is stretched on a rack of sententious self-interest, and ethical boundaries blurred way beyond anything of conscionable cartographical interest.

The watershed, according to Elson, was President Clinton's affair with Monica Lewinsky, and the fact that his administration, his family, and (to a large extent) his legacy survived relatively intact in the aftermath. Be that as it may, honor and trust continue to falter elsewhere. The police are under fire for institutional racism. Sports are under fire for the widespread use of performance-enhancing drugs. And the Catholic Church is under fire for sexually abusing minors.

The law itself has even gotten in on the act. At the Elizabeth Smart kidnapping trial in Salt Lake City, the attorney representing Brian David Mitchell—the homeless street preacher and self-proclaimed prophet who abducted, raped, and kept the fourteen-year-old Elizabeth captive for nine months (according to Smart's testimony he raped her pretty much every day over that period, by the way)—urged the sentencing judge to go easy on his client, on the grounds that "Ms. Smart overcame it. Survived it. Triumphed over it."

When the lawyers start whipping up that kind of tune, the dance could wind up anywhere.

Generation Me

I put it to Pinker, over lunch at the Harvard Faculty Club, that we've got a bit of a conundrum on our hands. On the one hand we have evidence that society is becoming less violent, while on the other there's evidence that it's getting more psychopathic.

He raises a good point. "Okay. Let's say that society *is* becoming more psychopathic," he counters. "That doesn't necessarily entail that there's going to be an upsurge in violence. The majority of psychopaths, as far as I understand it, are actually nonviolent. They inflict predominantly emotional, rather than physical, pain . . .

"Of course, if psychopathy really starts getting a toehold, then we might see a minimal increase in violence compared to what we would have seen, say, forty or fifty years ago. But what's probably more likely is that we'll start to detect a difference in the pattern of that violence. It might, for example, become more random. Or more instrumental.

"I think that society is going to have to get very psychopathic indeed for us to start living up to how we were, say, back in the Middle Ages. And, from a purely practical point of view, that level of manifestation is simply not attainable.

"It wouldn't surprise me one bit to find that subtle fluctuations in personality or interpersonal style have been occurring over the past few decades. But the mores and etiquette of modern civilization are far too deeply ingrained in us, far too embedded within our better natures, to be subverted by a swing, or what's probably more likely, a nudge, toward the dark side."

Pinker's right about psychopathy not being sustainable over the long term. As we saw with the aid of game theory in the previous chapter, it's a biological nonstarter. He's also right about changes in the motivation for violence. In a recent study by the Crime and Justice Centre at King's College, London, 120 convicted street robbers were asked, quite simply, why they did it. Their answers revealed a great deal about modern British street life. Kicks. Spur-of-the-moment impulses. Status. And financial gain. In precisely that order of importance. Exactly the kind of casual, callous behavior patterns one often sees in psychopaths.

So are we witnessing the rise of a sub-psychopathic minority for whom society doesn't exist? A new breed of individual with little or no conception of social norms, no respect for the feelings of others, and scant regard for the consequences of their actions? Might Pinker be right about those subtle fluctuations in modern personality structure—and a nefarious nudge to the dark side? If the results of a recent study by Sara Konrath and her team at the University of Michigan's Institute for Social Research are anything to go by, then the answer to these questions is yes.

In a survey that has so far tested fourteen thousand volunteers, Konrath has found that college students' self-reported empathy levels (as measured by the Interpersonal Reactivity Index*) have actually been in steady decline over the previous three decades—since the inauguration of the scale, in fact, back in 1979—and that a particularly pronounced slump has, it turns out, been observed over the past ten years.

"College kids today are about 40 percent lower in empathy than their counterparts of twenty or thirty years ago," Konrath reports.

More worrying still, according to Jean Twenge, a professor of psychology at San Diego State University, is that, during this same period, students' self-reported narcissism levels have, in contrast, gone in the other direction. They've shot through the roof.

* The Interpersonal Reactivity Index (IRI) is a standardized questionnaire containing such items as "I often have tender, concerned feelings for people less fortunate than me" and "I try to look at everybody's side of a disagreement before I make a decision."

"Many people see the current group of college students, sometimes called 'Generation Me,'" Konrath continues, "as one of the most self-centered, narcissistic, competitive, confident and individualistic in recent history."

Hardly surprising, then, that the former head of the British armed forces, Lord Dannatt, has recently championed the notion that recruits undergo a "moral education" as part of their basic training, so lacking are many in a basic, core value system.

"People haven't had the same exposure to traditional values which previous generations did," Dannatt articulates, "so we feel it's important people have a moral baseline."

Throw 'em in the army, they used to say of delinquents. Not anymore. They've got enough of 'em already.

Precisely *why* this downturn in social values should have come about is not entirely clear. A complex concatenation of environment, role models, and education is, as usual, doing the rounds. But the beginnings of an even more fundamental answer may lie in another study conducted by Jeffrey Zacks and his team at the Dynamic Cognition Laboratory, Washington University in St. Louis. With the aid of fMRI, Zacks and his coauthors peered deep inside the brains of a bunch of volunteers as they read stories. What they found provided an intriguing insight into the way our brain constructs our sense of self. Changes in characters' locations (e.g., "went out of the house into the street") were associated with increased activity in regions of the temporal lobes involved in spatial orientation and perception, while changes in the objects a character interacted with (e.g., "picked up a pencil") produced a similar increase in a region of the frontal lobes known to be important for controlling grasping motions. Most important of all, however, changes in a character's goal elicited increased activation in areas of the prefrontal cortex, damage to which results in impaired knowledge of the order and structure of planned, intentional action.

Imagining, it would seem, really *does* make it so. Whenever we read a story, our level of engagement with it is such that we "mentally simulate each new situation encountered in a narrative," according to lead researcher Nicole Speer. Our brains then interweave these newly encountered situations with knowledge and experience gleaned from our own lives, to create an organic mosaic of dynamic mental syntheses.

Reading a book carves brand new neural pathways into the ancient cortical bedrock of our brains. It transforms the way we see the world. Makes us, as Nicholas Carr puts it in his recent essay "The Dreams of Readers," "more alert to the inner lives of others." We become vampires without being bitten—in other words, more empathic. Books make us see in a way that casual immersion in the Internet, and the quicksilver virtual world it offers, doesn't.*

Guilty, But Not to Blame

Back in Montreal, Bob Hare and I down another whiskey. On the subject of empathy and perspective taking, we've been talking about the emergence of neurolaw, a developing subdiscipline born out of the increasingly greater interest the courts are now taking in cutting-edge neuroscience. The watershed study was published in 2002, and found that a functional polymorphism in a neurotransmitter-metabolizing gene predicts psychopathic behavior in men who were mistreated as children. The gene in question—termed, as mentioned previously, the "warrior gene" by the media—controls the production of an enzyme called monoamine oxidase A (MAOA), low levels of which had previously been linked with aggressive behavior in mice.

But Avshalom Caspi and Terrie Moffitt of the Institute of Psychia-

* According to a 2011 survey conducted by the U.K. charity the National Literacy Trust, one in three children between the ages of eleven and sixteen do not own a book, compared with one in ten in 2005. That equates, in today's Britain, to a total of around 4 million. Almost a fifth of the eighteen thousand children polled said they had never received a book as a present. And 12 percent said they had never been to a bookshop.

try, King's College, London, pushed the envelope further. In a trail-blazing study, which assessed children through adolescence into adulthood, they discovered a similar pattern in humans. Boys who are abused or neglected, and who possess a variation of the gene that codes for low levels of MAOA, are at an increased risk, as they get older, of turning into violent psychopaths. On the other hand, those coming from a similarly dysfunctional background, but who produce more of the enzyme, rarely develop such problems.

The implications of the discovery have percolated into the courtroom, and could completely rewrite the fundamental rules of crime and punishment. Whether we're "good" or whether we're "bad" lies partly in our genes, and partly in our environment. But since we don't choose either, are we free to choose at all?

In 2006, Bradley Waldroup's defense attorney, Wylie Richardson, summoned Professor William Bernet, a forensic psychiatrist from Vanderbilt University in Nashville, Tennessee, to the witness stand. Bernet had quite a job on his hands. Waldroup stood accused of one of the most brutal and heinous crimes in Tennessee's history. Following a visit from his estranged wife, their four children, and his estranged wife's friend to his trailer home, Waldroup, in his own words, "snapped." He picked up his .22 rifle, proceeded to drill eight peremptory holes in the friend's back, and then sliced her head open with a machete. Turning the machete on his wife, he lopped off her finger, then stabbed and slashed her repeatedly before changing tack and beating her senseless with a shovel.

Miraculously, his wife survived. But her friend, unfortunately, didn't. Which meant that Waldroup, if found guilty, faced the death penalty.

Richardson had other ideas. "Is it true," he asked Bernet, "that the accused possesses the variation of the gene that codes for low levels of MAOA?"

"Yes," replied Bernet.

"Is it also true," continued Richardson, "that he was violently, and repeatedly, beaten by his parents as a child?"

"Yes," replied Bernet.

"Then to what extent is the man who stands before you completely responsible for his actions?" persisted Richardson. "To what extent has his free will been eroded by his genetic predisposition?"

It was a groundbreaking question—particularly for Bradley Waldroup, whose very existence, depending on the outcome, hung precariously in the balance.

It got an equally groundbreaking answer. Enough, thought the court, to absolve him of first-degree murder and find him guilty of voluntary manslaughter. Enough, it turned out, for history to be made. The science of behavioral genomics had commuted an otherwise certain death sentence to life.

The subject of neurolaw came up in the context of a wider discussion about the field of cultural neuroscience: the study of how societal values, practices, and beliefs shape, and are shaped by, genomic, neural, and psychological processes across multiple timescales and cultures. If society was becoming increasingly psychopathic, I wondered, was there a gene already at work out there churning out more psychopaths? Or was it a case, as Steven Pinker had elucidated in his "culture of dignity" argument, of customs and mores becoming ever more socialized until they end up second nature?

Hare suggests that it's probably a little of both: that psychopaths, right now, are on a bit of a roll, and that the more of a roll they get on, the more normative their behavior becomes. He points to the emergence of epigenetics—a hot new offshoot from the field of mainstream genetics, which, put simply, looks at changes in gene activity that don't actually involve structural alterations to the genetic code per se, but still get passed on to successive generations. These patterns of gene expression are governed by little "switches" that sit on top of the genome, and it's through tampering with these switches, rather than through intricate internal rewiring, that environmental factors like diet, stress, and even prenatal nutrition can have their say—can, like mischievous biological poltergeists, turn your genes on or off, and make their presence felt in ancestral rooms long ago inherited from their original owner-occupiers.

Hare tells me about a study conducted in Sweden back in the 1980s. In the first half of the nineteenth century, a remote area in the north of the country, Överkalix, was stricken by a sequence of poor and unpredictable harvests. Years of famine were duly interspersed with bumper years of plenty.

Sifting through the data from meticulous agricultural archives and then comparing it with data from subsequent national health records, scientists uncovered something truly remarkable: an epidemiological inheritance pattern that turned the science of genetics on its head.

The sons and the grandsons of men whose prepuberty years* just so happened to coincide with a time of famine had, it turned out, a decreased risk of dying from cardiovascular disease (such as stroke, high blood pressure, or coronary problems). On the other hand, however, the sons and grandsons of men whose prepubescence coincided with a bumper harvest had, in contrast, an increased risk of succumbing to diabetes-related illnesses.

It was incredible. Through no direct agency of their own, successive generations of sons and grandsons had had their cardiovascular and endocrinological futures underwritten by the random ecological exigencies of an ancestral time long gone. Before they were even born.

"So is it possible," I ask, in an attempt to draw everything together—Pinker and his cultural arbiters, Boddy and his corporate Attilas, and the whole epigenetics shebang—"that psychopaths have rolled the dice, and that, over time, more and more of us are now rolling it with them?"

Hare orders another couple of shots.

"Not only that," he says. "But, over time, as you say, if the hand of epigenetics starts meddling behind the scenes, those dice will start to become more and more loaded. There's no doubt that there are

* More specifically, the Slow Growth Period (SGP)—the time immediately before the onset of puberty, when environmental factors have a larger impact on the body. For boys, this critical time window usually falls between the ages of nine and twelve.

elements of the psychopathic personality ideally suited to getting to the top. And once there, of course, they can start calling the tune to which others of their number are best suited to dance . . . Look at what's happened on Wall Street, for example . . . That's come from the top down. But, as it takes hold, it enables those best equipped to deal with such an environment, at lower levels of management, to start making their way up . . .

"There was a writer back in the sixties, Alan Harrington, who thought that the psychopath was evolution's next step. The next trick that natural selection has up its sleeve as society becomes faster and looser. Maybe he's right? There's no way of telling right now. But there's certainly some interesting work being done in genetics labs at the moment.

"Did I tell you about [this] paper which shows that people with high testosterone levels, and long alleles on their serotonin transporter genes, exhibit a suppressed amygdala response when faced with social dominance threats?

"That's a potential psychopath gene right there for you. You've got high aggression and low fear all rolled into one . . ."

Gary Gilmore's Eyes

I glance at my watch. It's a little after nine and the bar is filling up. As an amusing backdrop, "Gary Gilmore's Eyes" by the Adverts is playing—a post-punk ditty from yesteryear in which the singer muses on what it would be like to see through them. It's an interesting question—to which someone knows the answer. Following his execution, Gilmore had requested that his eyes be used for transplant purposes. Within hours of his death, in compliance with his wishes, two people received his corneas.

Gilmore, of course, is one of criminal history's superpsychopaths—one of that rare sub-breed of the species with all the dials on the mixing deck turned up to max. In January of 1977, the former shoe salesman was executed by a firing squad in the small and otherwise unremarkable town of Draper, Utah. The previous July, at a gas station a few miles up the road, he'd gunned down an attendant for reasons he wasn't quite sure of—and then he'd gone to see a movie with his girlfriend. The next day, as an encore, he'd shot a motel clerk in the head.

Six months later, after a final repast of hamburgers, eggs, and potatoes, he was facing the music in Utah State Prison. There were five in the firing squad. The prison warden tightened the leather straps around Gilmore's head and chest and fixed a circular cloth target over his heart. He then withdrew from the execution chamber and pressed his face against the cool, clear glass of the observation suite.

There was nothing now for Gilmore save a miraculous last-minute reprieve. And neither miracles nor reprieves were all that common in Draper at the time. Besides, a couple of months earlier, Gilmore had dropped his appeal. He actually, or so he'd told his attorney, wanted to die.

It was eight in the morning when the firing squad picked up their rifles. Before (as is tradition) pulling a black corduroy hood over his head, the warden (as is also tradition) asked Gilmore if there were any last words. Gilmore stared straight ahead, his eyes colder than a Great White's, as the inaudible thunder of death rumbled across his soul.

"Let's do it," he said.

As the song plays out, I turn, somewhat pensively, to Hare. "I wonder what it *would* be like to see through Gilmore's eyes," I say. "I mean—for real. If someone could turn you into a psychopath for an hour, would you do it?"

He laughs. "Maybe I would *now*," he drawls. "At my age. But they'd have to get the keys to my BMW off me first!"

We finish our drinks and go our separate ways. The song has got me thinking, and I prowl around the streets of Old Montreal with a crazy idea buzzing around my head. What about that study by Ahmed Karim: the one in which he'd made people better liars by zapping their moral decision-making regions—their anterior prefrontal cortices—with transcranial magnetic stimulation? If you can turn one of the dials up higher, then why not several others?

Magnetic Personality

Transcranial magnetic stimulation (or TMS) was first developed by Dr. Anthony Barker and his colleagues at the University of Sheffield, in 1985. But it's got a longer history than that. In fact, the science behind the electrical stimulation of the nerves and muscles has been around since the 1780s, some two hundred years before Barker plugged himself in, when the Italian anatomist and physician Luigi Galvani and his compatriot Alessandro Volta first discovered, with the aid of a simple electrical generator and a pair of severed frog's legs, that the nerves were not water pipes as René Descartes had conjectured, but electrical conductors that ferried information round the nervous system.

Since then, things have come a long way. The inaugural application of TMS by Barker and his team comprised an elementary demonstration of the conduction of nerve impulses from the motor cortex to the spinal cord by stimulating simple muscle contractions. Nowadays it's a different story—and TMS has widespread practical uses, in both a diagnostic and a therapeutic capacity, across a variety of neurological and psychiatric conditions, ranging from depression and migraine to strokes and Parkinson's disease.

The basic premise of TMS is that the brain operates using electrical signals, and that, as with any such system, it's possible to modify the way it works by altering its electrical environment. Standard equipment consists of a powerful electromagnet placed on the scalp that generates steady magnetic field pulses at specific preselected frequencies and a plastic-enclosed coil to focus those magnetic pulses down through the surface of the skull onto specially targeted, discretely segregated brain regions, thus stimulating the underlying cortex.

Now, one of the things that we know about psychopaths is that the light switches of their brains aren't wired up in quite the same way as the rest of ours—and that one area particularly affected is the amygdala, a peanut-sized structure located right at the very center of the circuit board. The amygdala, as we've learned previously in this book, is the brain's emotion control tower. It polices our emotional airspace and is responsible for the way we feel about things. But in psychopaths, a section of this airspace, the part that corresponds to fear, is empty.

Using the light switch analogy, TMS may be thought of as a dimmer switch. As we process information, our brains generate small electrical signals. These signals not only pass through our nerves to work our muscles but also meander deep within our brains as ephemeral electrical data shoals, creating our thoughts, memories and feelings. TMS can alter the strength of these signals. By passing an electromagnetic current through precisely targeted areas of the cortex, we can turn these signals either up or down—either help these data shoals on their way or impede their progress.

Turn down the signals to the amygdala, of course, and, as Ahmed Karim and his colleagues at the University of Tübingen did, to the brain's morality neighborhood, and you're well on the way to giving someone a "psychopath makeover." Indeed, Liane Young and her team at MIT have since kicked things up a notch and demonstrated that applying TMS to the right temporoparietal junction—a specific neural zip code within that neighborhood—has significant effects not just on lying ability, but also on moral reasoning ability: in particular, ascribing intentionality to others' actions.

I pick up the phone to my old friend Andy McNab. He's on a weeklong spree in the desert when I call, roaring around Nevada on a Harley V-Rod Muscle.

"No helmets!" he booms.

"Hey, Andy," I say. "You up for a little challenge when you get back?"

"Course!" he yells. "What is it?"

"How about you and me go head-to-head in a test of cool in the lab? And I come out on top?"

Manic laughter.

"Love it," he says. "You're on! Slight problem though, Kev. How the fuck do you think you're going to pull that one off?"

"Simple," I say.

Special "Author" Service

For those of you who've been living in a cave for the past twenty years (with the possible exception of the Taliban), Andy McNab was arguably the most famous British soldier to have served in Her Majesty's Armed Forces until Prince Harry hung up his polo mallet at Eton back in 2005. During the First Gulf War, Andy commanded Bravo Two Zero, an eight-man Special Forces patrol that was assigned the task of gathering intelligence on underground communication links between Baghdad and northwest Iraq, and tracking and destroying Scud missile launchers along the Iraqi main supply route in the area.

But soon the boys had other fish to fry. A couple of days after insertion, the patrol was compromised by a goatherd tending his flock. And, in time-honored fashion, they beat it: 185 miles, across the desert, toward the Syrian border.

Only one of them made it. Three were killed, and the other four, including Andy, were picked up at various points along the way by the Iraqis. Suffice it to say that none of their captors were ever going to have their own talk shows . . . or make their mark in the annals of cosmetic surgery. It's generally accepted that there are better ways of putting a person at ease than by stubbing your cigarette out on their neck. And better ways of breaking and remodeling their jawline than with the sun-baked butt of an AK-47. Thanks to more advanced techniques back home in the U.K., Andy's mouth now packs more porcelain than all the bathrooms in Buckingham Palace put together. He should know. In 1991, he went there to collect the Distinguished Service Medal from the Queen.

The medal was just the beginning. In 1993, in a book that bore their name, Andy told the story of the patrol, in all its gruesome detail, and defined, pretty much overnight, both the genre and the shape of the modern military memoir. In the words of the Special Air Service's (SAS) commanding officer at the time, the story of Bravo Two Zero "will remain in Regimental history forever." He wasn't kidding. In fact, it's now become part of a wider cultural history. And Andy has become a brand.

On a night flight to Sydney several years ago, we routed over Afghanistan. Down below, in the deep, dangerous darkness between the mountains of the Hindu Kush, I noticed, through the intermittent cloud cover, tiny pinpricks of light. What the hell were they? I wondered. The flickering campfires of age-old nomadic pastoralists? The secret hideouts of one-eyed Taliban warlords?

Right on cue, the pilot switched on the intercom. "For those of you on the right-hand side of the aircraft," he intoned, "you should just about be able to make out the laptops of the SAS as they bash out their latest bestsellers."

The intercom crackled off. And everyone laughed. Andy would've, too, if he'd been there. But I think we were flying over him at the time.

One of the first things you learn about Andy, and you learn it pretty quickly, is that he couldn't give a shit about anything. Nothing is sacred. And nothing remotely fazes him.

"I was a couple of days old when they found me," he explains as we meet for the first time at London Bridge station. "Just round the corner from here, in fact, on the steps of Guy's Hospital. Apparently I was wrapped in a Harrods bag."

"You're joking," I say. "Seriously?"

"Yeah," he says. "Straight up."

"Shit," I say. "Incredible. I had you down as more of a T.J.Maxx man, myself."

"You cheeky bastard!" he roars. "Nice one. I like it."

We've teamed up as part of a radio show I'm putting together for the BBC. The show is called *Extreme Persuasion*, and I'm interested in learning if certain psychopathic characteristics might come in handy in the SAS. Like, for instance, not giving a shit about anything.

I'm not disappointed. If you're thinking of joining, I'll tell you something for nothing. If you've got issues over your parentage, you're better off staying at home.

"One of the first things you notice on camp is the banter," Andy tells me. "It's pretty much constant. Everyone is always slagging everyone else off. Taking the piss. And like most things in the regiment, there's a good reason for that. If you're captured, you're taught to be the 'gray man.' To act tired and out of it. To give your interrogators the impression that you don't know shit. That you're of little use to them.

"Now, if [your captors are] any good, they'll start to look for weaknesses. They'll look for the tiniest of reactions—fleeting microexpressions, infinitesimal eye movements—that might give away your true mental state. And if they find anything, then take it from me: that's that, mate. It's game over. Put it this way. If you've got a problem with the size of your dick, an Iraqi interrogation suite probably isn't the best place to find out.

"So, in the regiment, everything's fair game. The slagging is purely functional. It's an efficient way of building up psychological immunity. It inoculates you against the kind of shit they can throw at you if you're captured. It's the right kind of wrong, if you see what I mean. Plus, you know, there's nothing like a good windup, really, is there?"

No, I guess not. But mental toughness isn't the only characteristic that Special Forces soldiers have in common with psychopaths. There's also fearlessness. A couple of years ago on a beautiful spring morning twelve thousand feet above Sydney's Bondi Beach, I performed my first free-fall skydive. The night before, somewhat the worse for wear in one of the city's waterfront bars, I texted Andy for some last-minute advice.

"Keep your eyes open. And your arse shut," came the reply.

I did. Just. But performing the same feat at night, in the theater of war, over a raging ocean from twice the altitude and carrying two hundred pounds of equipment, is a completely different ball game altogether. And if that's not enough, there's also the piss-taking to contend with. Even at thirty thousand feet, the party's going strong.

"We used to have a laugh," Andy recalls. "Mess about. You know, we'd throw the equipment out ahead of us and see if we could catch up with it. Or on the way down, we'd grab each other from behind in a bear hug and play chicken—see who'd be the first to peel off and pull the cord. It was all good fun."

Er, right. If you say so, Andy. But what wasn't much fun was the killing. I ask Andy whether he ever felt any regret over anything he'd done. Over the lives he'd taken on his numerous secret missions around the world.

"No," he replies matter-of-factly, his arctic blue eyes (just for the record, there are eyes behind the bars you see in the photos) showing not the slightest trace of emotion. "You seriously don't think twice about it. When you're in a hostile situation, the primary objective is to pull the trigger before the other guy pulls the trigger. And when you pull it, you move on. Simple as that. Why stand there, dwelling on what you've done? Go down that route and chances are the last thing that goes through your head will be a bullet from an M16.

"The regiment's motto is 'Who Dares Wins.' But sometimes it can be shortened to 'Fuck It.'"

Bonds of Detachment

Put like this, it's not difficult to envisage how such pathological poise, such conscienceless composure, might actually come in handy in certain circumstances—how it might, at times, actually be construed as adaptive. One of Andy's compatriots, Colin Rogers, a former member of the famed SAS assault team who, in Operation Nimrod back in 1980, tapped gently on the windows of the Iranian embassy in London, echoes his old mate's sentiments. Taking out a terrorist in the dust and the fire and the rubble that comprise the usual architectural legacy of covert explosive entry is not something that Special Forces soldiers tend to deliberate over too much—especially when you've got a state-of-the-art Heckler & Koch MP5 submachine gun, which fires 800 rounds a minute, slung over your shoulder and when the margin of error is often on the order of millimeters. Get a clear shot and you go for it. You focus. Stay calm. And coolly squeeze the trigger. Hesitation is not an option.

The trick, it would appear, is being fireproof. It's being able to perform not just in the heat of the moment, but simply in the moment. It's about not feeling hot in the first place.

"You're pumped up. Of course you are," Colin tells me, in his East End local pub that's clearly seen better days. "But this is something you've been training for for years. Six, seven hours a day. It's like driving. No journey is exactly the same. But you can cope pretty well with most eventualities. Your reactions become automatic. You use your judgment, yes. But even that's a product of the training. It's difficult to describe if you haven't actually been there. It's as if you've got a heightened sense of awareness of everything that's going on around you. Like the opposite of being sloshed. But at the same time you're also sort of outside the situation. Like you're watching it on film."

He's right. And not just when it comes to storming embassies. Recall the words of the neurosurgeon in the previous chapter? "An intoxication that sharpens rather than dulls the senses" is how he described it: the mind-set he enters before taking on a difficult operation. In fact, in any kind of crisis, the most effective individuals are often those who stay calm—who are able to respond to the exigencies of the moment while at the same time maintaining the requisite degree of detachment.

Consider the following, for example: an extract from an interview I did with a U.S. Special Forces instructor, about the caliber of soldier that eventually, after one of the most grueling physical and psychological selection procedures in the world, makes it into the Navy SEALs. The guys who took out Bin Laden.

> We did everything we could to break this guy. In fact, to be honest, we worked a bit harder on him than on the others. It got to be a bit of a challenge with us. Plus we sort of knew, deep down, he could take it. He was orphaned at eleven, but slipped through the net, looking after his younger brother and sister by living on his wits.

Stealing. Wheeling. Dealing. You know, that kind of thing. Then, when he was sixteen, he beat someone up so bad they went into a coma. And he got taken in.

White noise. Sleep deprivation. Sensory deprivation. Water. Stress positions. The lot. We threw everything at him. Finally, after forty-eight hours, I removed the blindfold, put my face within a few inches of his, and yelled:

"Is there anything you want to tell me?"

To my surprise, and I've got to say, disappointment—like I said, this guy was as hard as nails, and by this stage we were actually willing him to pass—he said yes. There was something he wanted to say.

"What is it?" I asked.

"You want to cut down on the garlic, dude," he said.

It was the only time, in fifteen years as an instructor, that I let my guard slip. Just for a second, a split second, I smiled. I couldn't help it. I actually admired this guy. And you know what? Even in the disgusting, fucked-up state he was in, the son of a bitch saw it. He saw it! He called me back closer to him. And there was a look of sheer, I don't know, defiance—who knows what it was?—in his eyes.

"Game over," he whispered in my ear. "You've failed."

What? I was meant to be saying that to him! It was then that we realized he was one of what we call the "unbreakables." The toughest of the tough . . .

But he was a ruthless fucking bastard. And if he DID have a conscience I never saw it. He was cold as ice. At either end of a weapon. Which actually, in this line of work, isn't always a bad thing.

McNab in the Lab

True to his word, Andy rocks up to the Centre for Brain Science at the University of Essex one bitterly cold December morning, and we're met at the door by the man who, for the next couple of hours or so, is going to be our tormentor. Dr. Nick Cooper is one of the world's leading exponents of TMS. And, from the look of him this morning, you could be forgiven for thinking he's done most of the work on himself.

Nick ushers us into the lab. The first things we notice are two high-backed leather chairs, side by side. And, next to them, the world's biggest roll of industrial paper toweling. I know what the toweling is for: it's to mop up the excess conductance gel that helps the EEG electrodes, which Nick is going to attach in a minute, pick up the signals from deep within our brains. Andy, on the other hand, has just his imagination to go on.

"Christ," he says, pointing at the stack. "If that's the size of the toilet paper, then I'm out of here!"

Nick shows us over to the chairs, and straps us in. He wires us up to heart-rate monitors, EEG recording equipment, and galvanic skin response (GSR) measures, which assess stress levels as a function of electrodermal activity. By the time he's finished, the pair of us look like we're trapped inside a giant telecom junction box. The gel for the electrodes feels cold against my scalp. But Andy isn't complaining. He's finally figured out what the king-sized TP roll is for.

Directly in front of us, about ten feet away on the wall, is a massive video screen. Nick flips a switch, which makes it crackle into life. Then he goes into white-coat mode. Ambient music wafts around the room. A silky, twilit lake ripples in front of our eyes.

"Bloody hell," says Andy. "It's like an ad for incontinence pads!"

"Okay," says Nick. "Listen up. Right now, on the screen in front of you, you can see a tranquil, restful scene which is presently being accompanied by quiet, relaxing music. This is to establish baseline physiological readings from which we can measure subsequent arousal levels.

"But at an undisclosed moment sometime within the next sixty seconds, the image you see at the present time will change and images of a different nature will appear on the screen. These images will be violent. And nauseating. And of a graphic and disturbing nature.

"As you view these images, changes in your heart rate, skin conductance, and EEG activity will be monitored, and compared with the resting levels that are currently being recorded. Any questions?"

Andy and I shake our heads.

"Happy?"

We nod.

"Okay," says Nick. "Let's get the show on the road."

He disappears behind us, leaving Andy and me merrily soaking up the incontinence ad. Results reveal later that, at this point, as we wait for something to happen, our physiological output readings are actually pretty similar. Both my and Andy's pulse rates are significantly higher than our normal resting levels in anticipation of what's to come.

But when Nick pulls the lever, or whatever it is that initiates the change of scene, an override switch flips somewhere in Andy's brain. And the ice-cold SAS soldier suddenly swings into action. As vivid, florid images of dismemberment, mutilation, torture, and execution flash up on the screen in front of us (so vivid, in fact, that Andy later confesses to actually being able to "smell" the blood: a "kind of sickly sweet smell that you never, ever forget"), accompanied, not by the ambient spa music of before, but by blaring sirens and hissing white noise, his physiological readings start slipping into reverse. His pulse rate begins to slow. His GSR begins to drop. And his EEG to quickly and dramatically attenuate. In fact, by the time the show is over, all three of Andy's physiological output measures are pooling *below* his baseline.

Nick has seen nothing like it. "It's almost as if he was gearing himself up for the challenge," he says. "And then, when the challenge eventually presented itself, his brain suddenly responded by injecting liquid nitrogen into his veins. Suddenly implemented a blanket neural cull of all surplus feral emotion. Suddenly locked down into a hypnotically deep Code Red of extreme and ruthless focus."

He shakes his head, nonplussed. "If I hadn't recorded those readings myself, I'm not sure I would have believed them," he continues. "Okay, I've never tested Special Forces before. And maybe you'd expect a slight attenuation in response. But this guy was in total and utter control of the situation. So tuned in, it looked like he'd completely tuned out."

Just as Bob Hare had found: the data was so freakish you really had to wonder who it came from.

Back at the ranch, and the news, needless to say, isn't nearly as cool. My physiological output readings have gone through the roof. Exactly like Andy's, they'd weighed in well above baseline as I'd waited for the carnage to commence. But that's where the similarity had ended. Rather than go down in the heat of battle, in the midst of the blood and guts mine had appreciated exponentially.

"At least it shows that the equipment is working properly," comments Nick. "And that you're a normal human being."

We look across at Andy, who's chatting up a bunch of Nick's Ph.D. students over by a bank of monitors. God knows what they make of him. They've just analyzed his data, and the electrode gel has done such a number on his hair he looks like Don King in a wind tunnel.

I, on the other hand, am still getting over the shock of some of the images. I feel sick. And jittery. And a little unsteady on my feet. I might, as Nick had pointed out, have shown up on the radar screens as normal. The needles and dials may well have vouched for my sanity. But I certainly don't feel normal as I cower in the corner of a beeping, flickering cubicle, poring over the data on a computer screen.

The difference in profiles is embarrassing. Whereas my EEG reading is a fair approximation of the New York skyline—a histographical cityscape of sheer, sharp, mathematical loft apartments—Andy's is like a top-end, low-rise golfing resort on one of those beautifully manicured islands in the middle of the Indian Ocean. Uniform. And compact. And insanely, uncannily symmetrical.

"Does make you wonder, doesn't it?" I turn to Nick. "What normal really is?"

He shrugs and resets the computer.

"Maybe you're about to find out," he says.

Make Me a Psychopath

All done, Andy is off to a luxury hotel in the country—where I'll be joining him later for a debrief. But that's only after I've run the gauntlet again, in phase II of the experiment. In which, with the aid of a "psychopath makeover," I'll have another crack of the whip. Another bash at the mayhem, carnage, and gore. Only this time, quite literally, with a completely different head on—thanks to the same kind of treatment that Ahmed Karim and Liane Young dished out in their moral processing experiments: a dose of TMS.

"The makeover does wear off, doesn't it?" Andy laughs. "'Cuz the hotel won't want two psychos propping up the bar."

"The effects of the treatment should wear off within half an hour," Nick elucidates, steering me over to a specially calibrated dentist's chair, complete with headrest, chin rest, and face straps. "Think of TMS as an electromagnetic comb, and brain cells—neurons—as hairs. All TMS does is comb those hairs in a particular direction, creating a temporary neural hairstyle. Which, like any new hairstyle, if you don't maintain it, quickly goes back to normal of its own accord."

Andy's face is a picture. What the hell is this? A lab or a salon?

Nick sits me down in the sinister-looking chair and pats me, a little too reassuringly for my liking, on the shoulder. By the time he's finished strapping and bolting me in, I look like Hannibal Lecter at LensCrafters. He positions the TMS coils, which resemble the handle part of a giant pair of scissors, over the middle section of my skull, and turns on the machine.

Instantly, it feels as if there's a geeky homunculus miner buried deep inside my head, tapping away with a rock hammer. I wouldn't say it was painful, but you wouldn't want him to be just clocking on—to be just at the start of his neuro-minerological shift.

"That's the electromagnetic induction passing down your trigeminal nerve," Nick explains. "It's one of the nerves responsible for sensation in the face, and for certain motor functions like biting, chewing, and swallowing. You can probably feel it going through your back teeth, right?"

"Right." I nod.

"What I'm actually trying to find," he continues, "is the specific part of your motor cortex responsible for the movement of the little finger of your right hand. Once we've pinpointed that, I can then use it as a kind of base camp, if you like, from which to plot the coordinates of the brain regions we're really interested in. Your amygdala, and your moral reasoning area."

"Well, you'd better get on with it," I mutter. "Because much more of this, and I'm going to end up strangling you."

Nick smiles.

"Blimey," he says. "It must be working already."

Sure enough, after about twenty seconds, I feel an involuntary twitch exactly where Nick has predicted. Weak, at first. Then gradually getting stronger. Pretty soon my right pinkie is really ripping it up. It's not the most comfortable feeling in the world—sitting strapped in a chair, in a dimly lit chamber, knowing that you don't have any control over the actions your body is performing. It's creepy. Demeaning. Disorienting . . . and kind of puts a downer on the whole free will thing, just a tad. My only hope is that Nick isn't in the mood to start clowning around. With the piece of gear he's waving about, he could have me doing cartwheels round the lab.

"Okay," he says. "We now know the location of the areas we need to target. So let's get started."

My little finger stops moving as he repositions his spooky neurological wand in the force field above my head. It's then just a matter of sitting there for a while as my dorsolateral prefrontal cortex and right temporoparietal junction get an electromagnetic comb-over. TMS can't penetrate far enough into the brain to reach the emotion and moral-reasoning precincts directly. But by damping down or turning up the regions of the cerebral cortex that have links with such areas, it can simulate the effects of deeper, more incursive influence.

It isn't long before I start to notice a fuzzier, more pervasive, more existential difference. Prior to the experiment, I'd been curious about the timescale: how long it would take me to begin to feel the rush.

Now I had the answer: about ten to fifteen minutes. The same amount of time, I guess, that it would take most people to get a buzz out of a beer or a glass of wine.

The effects aren't entirely dissimilar. An easy, airy confidence. A transcendental loosening of inhibition. The inchoate stirrings of a subjective moral swagger: the encroaching, and somehow strangely spiritual, realization that hell, who gives a shit, anyway?

There is, however, one notable exception. One glaring, unmistakable difference between this and the effects of alcohol. The lack of attendant sluggishness. The perseveration—in fact, I'd even say enhancement—of attentional acuity and sharpness. An insuperable feeling of heightened, polished awareness. Sure, my conscience certainly feels like it's been spiked with moral Rohypnol, my anxieties drowned with a half dozen shots of transcranial magnetic Jack Daniel's. But, at the same time, my whole way of being feels like it's been sumptuously spring-cleaned with light. My soul, or whatever you want to call it, immersed in a spiritual dishwasher.

So this, I think to myself, is how it feels to be a psychopath. To see through Gary Gilmore's eyes. To cruise through life knowing that no matter what you say or do, guilt, remorse, shame, pity, fear—all those familiar, everyday warning signals that might normally light up on your psychological dashboard—no longer trouble you.

I suddenly get a flash of insight. We talk about gender. We talk about class. We talk about color. And intelligence. And creed. But the most fundamental difference between one individual and another must surely be that of the presence, or absence, of conscience. Conscience is what hurts when everything else feels good. But what if it's as tough as old boots? What if one's conscience has an infinite, unlimited pain threshold and doesn't bat an eye when others are screaming in agony?

Ahem. Even more important: will my prosthetic psychopath implants make me cooler than Andy McNab?

Back in the chair, wired up to the counters and bleepers, I sit through the horror show again: the images modified, so as to avoid habituation. This time, however, it's a completely different story. "I know the guy before me found these images nauseating," I hear myself saying. "But actually, to be honest, this time round I'm finding it hard to suppress a smile."

The lines and squiggles corroborate my confession. Whereas previously, such was my level of arousal that it was pretty much a minor miracle that the state-of-the-art EEG printer hadn't blown up and burst into flames, my brain activity after the psychopath makeover is significantly reduced. Perhaps not quite as genteelly undulating as Andy's. But getting there, certainly. Not a New York skyscraper in sight.

It's a similar story when it comes to heart rate and skin conductance. In fact, in the case of the latter, I actually eclipse Andy's reading.

"Does that mean it's official?" I ask Nick, as we scrutinize the figures. "Can I legitimately claim to be cooler than Andy McNab?"

He shrugs. "I suppose," he says. "For now, anyway. But you'd better make the most of it while you can. You've got a quarter of an hour. Max."

I shake my head. Already I sense the magic wearing off. The electromagnetic sorcery starting to wane. I feel, for instance, considerably more married than I did a bit earlier—and considerably less inclined to go up to Nick's research assistant and ask her out for a drink. Instead, I go with Nick—to the student bar—and beat my previous best on Gran Turismo* out of sight. I floor it all the way round. But so what—it's only a game, isn't it?

"I wouldn't want to be with you in a real car at the moment," says Nick. "You're definitely still a bit ballsy."

It feels great. Not quite as good as before, perhaps, when we were in the lab. Not quite as . . . I don't know . . . "impregnable." But up there, for sure. Life seems full of possibility, my psychological horizons much broader. Why shouldn't I piss off to Glasgow this weekend for my buddy's stag party, instead of dragging myself over to Dublin to help my wife put her mother in a nursing home? Why not just do the

*Gran Turismo is a popular racing simulation video game.

opposite of what I normally would—and to hell with what people think? I mean, what's the worst that can happen? This time next year, this time next week even, it would all be forgotten. Who Dares Wins, right?

I take a couple of quid from the table next to ours—left as a tip, but who's going to know?—and try my luck on another couple of machines. I get to $100,000 on "Who Wants to Be a Millionaire," but crash and burn because I refuse to go 50-50. You couldn't have staged it better if you'd tried. I'm positive *American Psycho* is set in L.A., and nonchalantly push the button despite Nick's reservations.

It's New York.

"I thought you'd at least get that one," he laughs.

Then things start to change—pretty quickly, as it happens. Gran Turismo the second time around is a disappointment. I'm suddenly more cautious, and finish way down the field. Not only that, I notice a security camera in the corner and think about the tip I've just pocketed. To be on the safe side, I decide to pay it back.

Nick looks at his watch. I know what's coming—he doesn't need to tell me.

"Still cooler than McNab?"

I smile, and swig my beer. Psychopaths. They never stick around for long. As soon as the party's over, they're moving on to the next one—with scant regard for the future, and even less for the past.

And this one—the one, I guess, that was *me* for twenty minutes—was no exception. He'd had his fun. And got a free drink out of it. But now that the experiment was history, he was suddenly on his way, hitting the road and heading out of town.

Hopefully, quite some distance away.

I certainly didn't want him showing up in the hotel bar later, where I was meeting Andy. They'd either get on great. Or wouldn't get on at all.

To be perfectly honest, I didn't know which was scarier.

SIX

THE SEVEN DEADLY WINS

> Sentiment is a chemical aberration found on the losing side.
> —SHERLOCK HOLMES

Crossing the Border

The joke is that it's harder to get into Broadmoor than out. But it isn't. A joke, that is.

"Got anything sharp?" the woman at reception barks as I deposit the entire contents of my briefcase—laptop, phone, pens, yes, my trusty Glock 17 pistol—into a Plexiglas-fronted locker in the entrance hall.

"Only my wit," I reply, parodying Oscar Wilde's onetime comment to a U.S. customs official.

The receptionist isn't a fan. Either of me or, it appears, of Oscar.

"It's not that sharp, Sonny," she shoots back. "Now place the index finger of your right hand here and look up at the camera."

Once you pass through border control at Broadmoor, you're immediately ushered into a tiny security airlock, a temporary glass-walled holding cell between reception and the hospital building proper, while the person you're visiting gets buzzed by reception and makes their way over to meet you.

It's an unnerving, claustrophobic wait. As I flip through the magazines, I remind myself why I'm here—an e-mail I'd received after launching the Great British Psychopath Survey. The survey is unique: the first of its kind to assess the prevalence of psychopathic traits within an entire national workforce. Participants were directed onto my website, where they completed the Levenson Self-Report Psychopathy Scale and were then given their score. But that wasn't all. They also entered their employment details. What would turn out to be the U.K.'s most psychopathic profession? I wanted to know. And, for that matter, its least? The results, revealed below, certainly make interesting reading—especially if you're partial to a sermon or two on a Sunday.

+ PSYCHOPATHY	– PSYCHOPATHY
1. CEO	1. Care Aide
2. Lawyer	2. Nurse
3. Media (TV/Radio)	3. Therapist
4. Salesperson	4. Craftsperson
5. Surgeon	5. Beautician/Stylist
6. Journalist	6. Charity Worker
7. Police Officer	7. Teacher
8. Clergyperson	8. Creative Artist
9. Chef	9. Doctor
10. Civil Servant	10. Accountant

But a couple of weeks later the following appeared in my in-box, from one of the survey's respondents. He's a barrister by trade—indeed, one of the U.K.'s finest—who'd posted a score that certainly got my attention. Yet, to him, it was nothing unusual. No big deal whatsoever:

"I realized from quite early on in my childhood that I saw things differently from other people," he wrote. "But, more often than not, it's helped me in my life. Psychopathy (if that's what you want to call it) is like a medicine for modern times. If you take it in moderation it can prove extremely beneficial. It can alleviate a lot of existential ailments that we would otherwise fall victim to because our fragile psychological immune systems just aren't up to the job of protecting us. But if you take too much of it, if you overdose on it, then there can, as is the case with all medicines, be some rather unpleasant side effects."

The e-mail had got me thinking. Might this eminent criminal defense lawyer have a point? Was psychopathy a "medicine for modern times"? Could taking it in moderation, twiddling those dials a little to the right on our respective psychopath mixing desks—at certain times, in certain specific contexts—actually be good for us?

It was an interesting possibility. And moreover, it made a lot of intuitive sense. Let's take a look at those dials for a moment: ruthlessness, charm, focus, mental toughness, fearlessness, mindfulness (living in the moment), and action. Who wouldn't, at certain points in their lives, benefit from kicking one or two of them up a notch? What was important was being able to turn them back down.

I decided to put the theory to the test—not quite to destruction, but it was sure as hell going to be close. I had a visit to several hospitals coming up, to interview some colleagues. But what if I went on the wards? What if, as well as meeting the doctors, I talked with some of the patients? Presented them with problems from normal, everyday life, the usual stuff we moan about at happy hour, and see what their take on it was? See what suggestions they happened to come up with? Up until now, it had seemed like a good idea.

"Professor Dutton?" My train of thought is broken and I look up to see a blond guy in his mid-thirties peering round the door at me. "Hi, I'm Richard Blake—one of the team leaders at the Paddock Centre. Welcome to Broadmoor! Shall I take you over?"

As we weigh anchor and begin charting our course deeper and deeper into the mazy, medicinal bowels of the hospital, through a series of interconnecting corridors and no-man's-land antechambers just like the one we started out from—"security air bubbles," as Richard calls them: the golden rule in Broadmoor is never open any door in front of you before first making sure that the one behind is locked—he tells me, in a little bit more detail, about where we're going.

The Paddock Centre is an enclosed, highly specialized personality-disorder directorate comprising six twelve-bedded wards. Around

20 percent of the patients housed there at any one time are what you might call "pure" psychopaths, and these are confined to two dedicated wards specifically assigned to their treatment and continual assessment: the Dangerous and Severe Personality Disorder (DSPD) wards. The rest present with so-called cluster disorders: clinically significant psychopathic traits (as evidenced by moderately high PCL-R scores) accompanied by the presence of core supplementary traits typically associated with other certifiable personality disorders—borderline, paranoid, and narcissistic, for example. Or, alternatively, traits more indicative of primary psychotic symptomatology, delusions and hallucinations being the obvious cases in point.

Suddenly reality dawns. This is no drop-in center for the mocha-sipping worried well I'm about to enter. This is the lair, the conscienceless inner sanctum, of the Chianti-swilling unworried unwell—the preserve of some of the most sinister neurochemistry in the business, where brain states, quite literally, can teeter on a knife-edge. The Yorkshire Ripper is in here. So is the Stockwell Strangler. It's one of the most dangerous buildings on earth.

"Er, I am going to be all right, aren't I, Richard?" I squeak as we suddenly emerge to the right of a large open-air enclosure, topped off with some distinctly uncooperative razor wire.

He grins. "You'll be fine," he says. "Actually, trouble on the DSPD wards is relatively rare. Psychopathic violence is predominantly instrumental, a direct means to a specific end. Which means, in an environment like this, that it's largely preventable. And, in the event that something does kick off, easily contained. It's on the psychotic wards that things are less predictable.

"In fact, even when compared with other personality disorders, psychopaths are easier to deal with. For some reason, they tend to respond better to daily activities than, say, borderlines or paranoids. Maybe it's because of their low boredom threshold: they like to keep themselves amused.

"Besides," he adds, with just the merest hint of reproach, "it's a bit late to turn back now, isn't it?"

Getting to Know the Locals

"We are the evil elite," says Danny as he slams in his second goal for Chelsea, a belter of a header from right smack on the edge of the six-yard area. "Don't glamorize us. But, at the same time, don't go the other way and start dehumanizing us, either."

He shoots me a glance from behind his Nintendo Wii. Things are going well both on and off the field. Chelsea are up 2–0 against Manchester United—and I'm sitting watching, in the company of a bunch of psychopaths, feet up on a table, in the corner of one of Broadmoor's ultra sequestered DSPD wards.

The vibe on the ward isn't what I'm expecting. My first impression is of an extremely well-appointed college dormitory. All blond, clean-shaven wood. Voluminous, freshly squeezed light. And mathematically defibrillated space. There's even a pool table, I notice. Which, today, unfortunately—it would've been nice to have clawed back my train fare—has a sheet over it.

Larry, a gray, bewhiskered, roly-poly kind of guy, who, in a Fair Isle sweater and beige stretch slacks, looks like everyone's favorite uncle—except that if you were planning on going out for the evening you'd be better off hiring Herod as a babysitter—takes a shine to me. He's had enough of the soccer.

"You know," he says, as he shakes my hand and silently impales me on his somnolent, moonlit stare, "they say I'm one of the most dangerous men in Broadmoor. Can you believe that? But I promise you I won't kill you. Here, let me show you around."

Larry escorts me to the far end of the ward, where we stop to take a peek inside his room. It's the same, pretty much, as any other single-occupancy room you might find in a hospital, though with a few more creature comforts. Like a computer, for instance. And desk space. And a raft of books and papers on the bed.

Sensing, perhaps, my inchoate curiosity, he draws a little closer. "I've been in for twenty years," he sibilates in my ear. "That's a hell of a lot of time to, er . . ."—he clears his throat, and smiles conspiratorially—". . . have on your hands. Are you with me?"

Next station down is the garden: a sunken gray-bricked patio affair about the size, give or take, of a tennis court, interspersed with benches and conifers. Verdict: "Gets a bit samey after two decades."

Fair enough. We then venture over to the opposite side of the ward—the layout of the place is symmetrical: six rooms on one side, six on the other, separated, in the middle, by a well-vacuumed ash-gray meridian—and drop in on Jamie.

"This guy's from Cambridge University," announces Larry, "and he's in the middle of writing a book on us."

Jamie stands up and heads us off at the door. It's a clear invitation for us to kindly retrace our steps. Which we do—rather quickly—back into the sanctuary of the ward. Jamie, it transpires, is a completely different proposition from Larry. A monster of a man at around six foot two, with brutal, char-grilled stubble and a piercing cobalt stare, he has the brooding, menacing, sub-satanic presence of the lone ultraviolent killer. The lumberjack shirt and shaved wrecking-ball head don't exactly help matters.

"So what's this book about, then?" he growls in a gangsterish Cockney whisper, wedged into the doorframe of his room, arms folded in front of him, left fist jammed like a ball hammer under his chin. "Same old bollocks, I suppose? Lock 'em up and throw away the key? You know, you've got no idea how vindictive that can sound at times. And, might I add, downright hurtful. Has he, Larry?"

Larry guffaws theatrically and clasps his hands to his heart in a Shakespearean display of angst. Jamie, meanwhile, dabs at imaginary tears.

This is great. Precisely what I came here for. Such stoic irreverence in the face of unremitting adversity is something, perhaps, we could all do with a little bit more of.

"You know what, Jamie?" I say, "I'm trying to do exactly the opposite. I happen to think that you guys have got something to teach us. A certain personality style that the rest of us can learn from. In moderation, of course. That's important. Like the way, just now, you shrugged off what people might think of you. In everyday life, there's a level on which that's actually quite healthy."

Jamie seems quite amused by the idea that I might be soliciting his advice. That the polarized vantage point of a psychopath might, in actual fact, offer some valuable perspective on the dilemmas of everyday life. But he's still a little circumspect.

"Are you saying that me and Captain McAllister here have just got too much of a good thing?" he sneers. "That the car's pretty cool but the driver's too fast for the road?"

It's an interesting analogy.

"Kind of," I say, "You interested in taking your foot off the gas and pulling over for a minute?"

Jamie's eyes narrow. "I ain't pulling over for no one," he shoots back. "But if you fancy a ride, hop in."

Back where we started, down at the other end of the ward, Chelsea are now four ahead of United. And Danny—who else?—has just been named Man of the Match.

"I see he hasn't killed you then," he says casually, throwing a quick glance in Captain McAllister's direction. "You going soft in your old age, Larry?"

I laugh. More than a little nervously, I realize. There's a healthy tinge of mania in my chortling. But Larry is dead serious.

"Hey," he says insistently. "You don't get it, do you, boy? I said I wouldn't kill you. And I didn't, right?"

Suddenly it dawns on me that Larry may not have been bluffing. May well have been exerting a touch more self-control than appearances might have suggested. And that my nervous discomfiture and attempt to laugh it off have, far from achieving that noble and laudable aim, succeeded in getting his back up.

"No, I do, Larry . . ." I pipe up. " . . . Get it. Really. Thanks, man. It's very much appreciated."

Jamie smiles. He obviously finds it funny. But from the precariously thin ice that I now appear to be skating on, it's no laughing matter at all. It's easy to forget that anything is possible with these guys. That there really are no limits. And that with no moral brake pads and a V12 amygdala, it doesn't take much for the car to spin off the road.

The curtain comes down on the soccer. And Danny zaps it off. He leans back in his chair.

"So, a book, eh?" he says.

"Yes," I say. "I'm interested in the way you guys solve problems."

Danny eyes me quizzically. "What kind of problems?" he asks.

"Everyday problems," I say. "You know, the kind most people have to deal with in their lives."

I glance at Larry and Jamie. "Mind if I give you an example?"

Danny looks at the clock. "Why not?" he sighs. "As long as it doesn't take longer than five years."

"I'll try to keep it brief," I say—and tell him about some friends of mine who were trying to sell their house.

Ruthlessness

How to get rid of an unwanted tenant? That was the question for Don and his wife, Fran, who'd just had Fran's elderly mother, Flo, move in with them. Flo had lived in her previous house for forty-seven years, and now that she no longer needed it, Don and Fran had put it on the market. Being in London, in an up-and-coming area, there was quite a bit of interest. But also a bit of a problem: the tenant, who wasn't exactly ecstatic at the prospect of hitting the road.

Don and Fran were pretty much at the end of their tether. They'd already lost out on one potential sale because he couldn't, or wouldn't, pack his bags. Another might prove disastrous. But how to get him out?

"I'm presuming we're not talking violence here," inquires Danny. "Right?"

"Right," I say. "We wouldn't want to end up inside, now would we?"

Danny gives me the finger. But the very fact that he asks such a question at all debunks the myth that violence, for psychopaths, is the only club in the bag.

"How about this, then?" rumbles Jamie. "With the old girl up at her in-laws', chances are the geezer's going to be alone in the house, yeah? So you pose as some bloke from the council, turn up at the door, and ask to speak to the owner. He answers and tells you the old dear ain't in. Okay, you say. Not a problem. But have you got a forwarding contact number for her? 'Cuz you need to speak to her urgently.

"By this stage he's getting kind of curious. What's up? he asks, a bit wary, like. Actually, you say, quite a lot. You've just been out front and taken a routine asbestos reading. And guess what? The level's so high it makes Chernobyl look like a health spa. The owner of the property needs to be contacted immediately. A structural survey has to be carried out. And anyone currently living at the address needs to vacate the premises until the council can give the all clear.

"'That should do the trick. With a bit of luck, before you can say 'slow, tortuous death from lung cancer,' the wanker will be straight out the door. Course, you could just change the locks when he nips out down the local, I s'pose. That'd be kind of funny. But the problem then is, you've still got all his gear. Which I guess is okay if you're planning on having a garage sale. I mean, you could even make a few quid out of the jerk and cover the cost of the locks . . .

"Me, though? I'd go the health-and-safety route, personally. Ha, stealth and safety, more like! You'd get shot of the bastard completely that way, I reckon. Plus, of course, he'd think you were doing him a favor."

Jamie's elegant, if rather unorthodox, solution to Don and Fran's stay-at-home tenant conundrum certainly had me beat. But there was, in my defense, a perfectly good reason for that, of course. I'm not a ruthless psychopath! The idea of getting the guy out so fast as to render him homeless and on the streets just simply hadn't occurred to me. It just hadn't flashed up on my radar. Neither, for that matter, had selling all his possessions to pay for the pleasure of locking him out of the house. And yet, as Jamie quite rightly pointed out, there are times in life when it's a case of the "least worst option." When sometimes, in order to achieve the desired, or most favorable, outcome, you've got to do what you've got to do.

But there's more. Interestingly, he argues it's actually the *right* thing to do: from an objective point of view, the ethical course of action.

"Why not turf the bastard out?" he asks. "I mean, think about it. You talk about 'doing the right thing.' But what's worse, from a moral perspective? Beating someone up who deserves it? Or beating yourself up, who doesn't? If you're a boxer, you do everything in your power to put the other guy away as soon as possible, right? So why are people prepared to tolerate ruthlessness in sport, but not in everyday life? What's the difference . . . ?

"The problem with a lot of people is that what they think is a virtue is actually a vice in disguise. It's much easier to convince yourself that you're reasonable and civilized than soft and weak, isn't it?"

"Good men sleep peacefully in their beds at night," George Orwell once pointed out, "because rough men stand ready to do violence on their behalf."

But perhaps, if one of the world's most dangerous psychopaths is to be believed, we could all do with a bit of a wake-up call.

Charm and Focus

Jamie's solution to Don and Fran's tenant problem unquestionably carries undertones of ruthlessness. Yet, as Danny's initial qualification of the dilemma quite clearly demonstrates—"I'm presuming we're not talking violence here, right?"—such ruthlessness need not be conspicuous. The more ingenious its deployment, the more creative the ruthlessness narrative, the greater your chances of pulling it off with impunity. The dagger of hard-nosed self-interest may be concealed, rather deftly, under a benevolent cloak of opaque, obfuscatory charm.

The psychopath's capacity for charm is, needless to say, well documented. As is their ability to focus and "get the job done." It goes without saying that it's a powerful, and smart, combination—and one that all of us could benefit from using.

Leslie has joined us and has a rather nice take on charm: "the ability to roll out a red carpet for those you cannot stand in order to fast-track them, as smoothly and efficiently as possible, in the direction you want them to go."

With his immaculately coiffured blond locks and his impeccable cut-glass accent, he looks and sounds like an authority. "People are as nice as you make them," he enunciates. "Which, of course, gives you a heck of a lot of power over them."

Leslie also has a good take on focus, especially when it comes to getting what you want. The master realized from a rather young age that what went on in his head obeyed a different set of operating principles compared with most—and he used that knowledge to his own inexorable advantage.

"When I was a kid at school, I tended to avoid fisticuffs," he tells me. "Same as I do as an adult. Rather like Jamie, I suppose."

Jamie smiles, with more than a hint of wry self-approbation.

"You see, I figured out pretty early on that, actually, the reason why people don't get their own way is because they often don't know themselves where that way leads. They get too caught up in the heat of the moment and temporarily go off track. At that point, the dynamic changes. That's when things become not just about getting what you want. But about being seen to get what you want. About winning.

"Jamie was talking about boxing there a minute ago. Well, I once heard a great quote from one of the top trainers. He said that if you climb into the ring hell-bent on knocking the other chap into the middle of next week, chances are you're going to come unstuck. But if, on the other hand, you concentrate on winning the fight, simply focus on doing your job, well, you might just knock him into the middle of next week anyway."

Leslie's words make perfect sense to me, and remind me of an encounter that took place several years ago—one in which vengeance and violence might easily have come into the equation, but where charm and focus won the day instead.

At six foot five and 250 pounds, Dai Griffiths was built more along the lines of a Greek restaurant than a Greek god. With twenty-three years' unbroken service in a certain British police force and a score on the PPI that probably placed him further along the psychopathic spectrum than most of the guys he arrested, he'd pretty much seen it all.

"Twenty percent of the people who come through those doors," he told me, gesturing to the entrance of the detention area, "take up 80 percent of our time." By which, of course, minus the fancy numerics, he meant that recidivists were a pain in the ass.

Recidivists, for example, like Iain Cracknell.

Cracknell was what you might call a career drunk. Regular as clockwork, on a Friday or Saturday night he'd be brought into the station with a golden future behind him. A bottle of Jack Daniel's, usually. And God knows how many beers.

What happened then was a routine so well choreographed it made *Swan Lake* look like a hoedown. First, Cracknell started to act "crazy." Next, a psychiatrist would be called in (as is required by law) to provide an assessment of his mental state. But by the time the shrink arrived, Cracknell—surprise, surprise—was back to normal. Drunk, for sure. But certainly not crazy. The psychiatrist would leave, mumbling something about police incompetence and uncivilized hours, and Cracknell, roaring with laughter, would be bundled into a cell to sleep it off. And then the same thing would happen the next time.

The problem with Cracknell seemed irresolvable. How to put an end to his interminable mind games? The trouble (as is the case with most repeat offenders) was that he knew the system better than anyone. And, of course, how to play it. Which meant you were left with a choice. You either didn't arrest him at all—or, if you did, you faced the consequences. Usually a blast from some seriously pissed-off psychiatrist.

And that, it appeared, was that.

Until one night, Griffiths has an idea. Having settled Cracknell into his customary weekend chambers and sent, as usual, for the duty psychiatrist, he makes his way along to the lost-property locker. A short time later, shuffling down the corridor in full clown's regalia—hair, rouge, nose, bells—he drops in on Cracknell again.

What, Griffiths inquires, would he like for his breakfast in the morning? Cracknell, to say the least, is incredulous. Sometimes, if he's lucky, he gets a glass of water. Not even a glass: a polystyrene cup. Now here he is getting the red carpet treatment. He can't believe his luck.

"And how would you like your eggs," Griffiths continues, "scrambled, poached, fried, or boiled?"

With the attention to detail of a top maître d', he makes a note of everything Cracknell asks for. Even the freshly squeezed orange juice. Then he leaves.

Ten minutes later, when he returns with the duty psychiatrist, he's back in uniform. "So," mutters the psychiatrist. "What seems to be the problem this time?"

Cracknell looks edgy.

"It's not me you should be talking to," he stammers. "It's him. You're not going to believe this. But, just before you got here, he was all done up in a clown's costume and asking me what I wanted for breakfast!"

The psychiatrist shoots Griffiths a suspicious glance. Griffiths just shrugs.

"Looks like we're in business," he says.

Dai Griffiths, take my word for it, isn't a man you want to get on the wrong side of in a hurry. Plenty of people have—and most of them ended up with a couple less teeth than they started with. He's not called the Dentist for nothing.

But Griffiths, quite clearly, has more than one string to his bow. He could easily have taught Cracknell a lesson. Drunks, as everyone knows, have "accidents." Bang into things. Pick up the odd bruise here and there. And yet he didn't. Instead, he went a different route entirely. He avoided the trap that Leslie had so eloquently warned of—the temptation to not just get what you want, but to be seen to get what you want: to show Cracknell who was boss behind closed doors on a personal, superfluous level—and focused, in contrast, on finding a solution that would resolve the dilemma once and for all, not just for himself but for his colleagues down the line. He concentrated on the issue at hand. Rolled out the red carpet. And eradicated the problem at source. Psychiatrists could put their feet up over the weekend.

Of course, the observation that charm, focus, and ruthlessness—three of the psychopath's most instantly recognizable calling cards—constitute, if you can juggle them, a blueprint for successful problem solving might not come as too great a surprise, perhaps. But that this triumvirate can also predispose—if the gods are really smiling on you—to inordinate, towering, long-term life success might well be a different matter.

Take Steve Jobs.

Jobs, commented the journalist John Arlidge shortly after Jobs's death, achieved his cult leader status "not just by being single-minded, driven, focused (he exuded, according to one former colleague, a 'blast-furnace intensity'), perfectionistic, uncompromising, and a total ball-breaker. All successful business leaders are like that, however much their highly paid PR honeys might try to tell us they are laid-back fellas, just like the rest of us . . ."

No. There was more to him than that. In addition, Arlidge notes, he had charisma. He had vision. He would, as the technology writer Walt Mossberg revealed, even at private viewings, drape a cloth over a product—some pristine new creation on a shiny boardroom table—and uncover it with a flourish.

Apple isn't the world's greatest techno-innovator. Nowhere near it, in fact. Rather, it excels at rehashing other people's ideas. It wasn't the first outfit to introduce a personal computer (IBM). Nor was it the first to introduce a smartphone (Nokia). Indeed, when it *has* gone down the innovation road, it's often screwed up. Anyone remember the Newton or the Power Mac G4 Cube?

But what Jobs did bring to the table was style. Sophistication. And timeless, technological charm. He rolled out the red carpet for consumers, from living rooms, offices, design studios, film sets—you name it—right to the doors of Apple stores the world over.

Mental Toughness

Apple's setbacks along the road to world domination (indeed, they were on the verge of going down the drain in the early days) serve as a cogent reminder of the pitfalls and stumbling blocks that await all of us in life. No one has it all their own way. Everyone, at some point or other, "leaves someone on the floor," as the Leonard Cohen song goes. And there's a pretty good chance that that someone—today, tomorrow, or at some other auspicious juncture down the line—is going to turn out to be you.

Psychopaths, lest Jamie and the boys have yet to disabuse you, have no problem whatsoever facilitating others' relationships with the floor. But they're also pretty handy when they find themselves on the receiving end—when fate takes a swing and they're the ones in the firing line. And such inner neural steel, such inestimable indifference in the face of life's misfortunes, is something that we could all, in one way or another, perhaps do with a little bit more of.

James Rilling, associate professor of anthropology at Emory University, has demonstrated this in the lab, and has discovered, in an iterated Prisoner's Dilemma task like the one we discussed in chapter 3, an odd yet uplifting paradox about the psychopath. Perhaps not surprisingly, psychopaths exhibit an enhanced propensity for "defection" under such conditions, which, in turn, precipitates elevated levels of belligerence, of opportunistic interpersonal aggression (encapsulated in the "cooperate-defect" dynamic) on the part of their opposite numbers.

Yet here's the deal. When the shoe's on the other foot, they simply aren't as bothered by such setbacks. Following the occurrence of these turnaround, "see how you like it" outcomes, in which those scoring high in psychopathy find their own attempts to cooperate unreciprocated, Rilling and his coworkers uncovered something interesting in their brains. Compared with their "nicer," more equitable fellow participants, the psychopaths exhibited significantly reduced activity in their amygdalae: a registered neural trademark of "turning the other cheek" . . . which can sometimes manifest itself in rather unusual ways.

"When we were kids," Jamie chimes in, "we'd have a competition. See who could get shot down the most times on a night out. You know, by girls, like. Although, in old McAllister's case here, we'd have had to have widened the field a bit."

Larry looks at me, nonplussed.

"Anyway, the bloke who'd got the most by the time the lights came on would get the next night out for free.

"Course, it was in your interest to rack up as many as possible, weren't it? A night on the town with everything taken care of by your mates? Aces! But the funny thing was, soon as you started to get a few under your belt, it actually got fucking harder. Soon as you realize that it actually means jack shit, you start getting cocky. You start mouthing off. And some of the birds start to buy it!"

Give rejection the finger and rejection gives it back.

Fearlessness

Jamie and company aren't the first to make the connection between fearlessness and mental toughness.

Lee Crust and Richard Keegan at the University of Lincoln, for example, have shown that the majority of life's risk takers tend to score higher on tests of general "mental toughness" than those who are risk averse, with scores on the challenge/openness-to-experience subscale being the single biggest predictor of physical risk taking, and scores on the confidence subscale being the biggest predictor of psychological risk taking. Both of which qualities psychopaths have in abundance.

Recall the words of Andy McNab in the previous chapter? You know there's a high chance of getting killed on a mission; you know there's a fair probability of being captured by enemy troops; you know there's a good possibility that you and your parachute will be swallowed up by waves the size of high-rises in some seething foreign ocean. But fuck it. You get on with it. That's what Special Forces soldiering is all about.

That members of Special Forces are both fearless and mentally tough (psychopathically so, it emerges, as the results of many of those I've tested bear witness) is beyond doubt. In fact, the instructors on the SAS's brutal, bestial selection course (which extends over a period of nine months, and which only a handful of candidates pass) are specifically on the lookout for such qualities—as some of the nightmares one endures on it attest to.

One example, recounted to me by a guy who came out on top, provides a pretty good insight into the kind of mental toughness that separates the men from the boys; that exemplifies the mind-set, the elite psychological makeup, of those who eventually prevail.

"It's not the violence that breaks you," he elucidates. "It's the threat of violence. That carcinogenic thought process that something terrible is going to happen. And that it's just around the corner."

He goes into details about one particular instance—which put me off fixing my car exhaust forever:

"Typically, by this stage, the candidate's exhausted . . . Then, the last thing he sees before we place the hood over his head is the two-ton truck. We lie him down on the ground, and as he lies there, he hears the sound of the truck getting closer. After thirty seconds or so, it's right there on top of him—the engine just inches away from his ear. We give it a good rev, and then the driver jumps out. He slams the door and walks away. The engine's still running. A little while later, from somewhere in the distance, someone asks if the handbrake's on. At this point, one of the team—who, unbeknownst to the guy in the hood, has been there all the time—gently starts to roll a spare tire onto his temple as he lies on the ground. You know, by hand. Gradually, he increases the pressure. Another member of the team revs the truck up a bit so it seems like it might be moving. After a few seconds of that, we take the tire away and remove the hood. Then we lay into him . . . It's not unusual for people to throw in the towel at that point."

I amuse the fellas—Danny, Larry, Jamie, and Leslie—with my own little taste of SAS selection, which I got while doing a TV pilot.

Shackled to the floor of a cold, dimly lit warehouse, I watched—in abject terror—as a forklift truck suspended a pallet of reinforced concrete several yards above my head . . . and then proceeded to lower it so that the sharp, rough-hewn base exerted a light, splintery pressure on my chest. It hovered there for about fifteen seconds before I heard the operator holler above the sinister, sibilant screeching of the hydraulics: "Shit, the mechanism's jammed. I can't shift it . . ."

In hindsight, after a hot bath, it soon became apparent that I'd been as safe as houses all along. In actual fact, the "reinforced concrete" hadn't been concrete at all. It was painted polystyrene. But needless to say, it would have been news to me at the time, And news to the Special Forces hopefuls who undergo such ordeals upon selection. In the moment, as I reported in *Split-Second Persuasion,* it's horribly real.

Jamie, however, is distinctly unimpressed. "But even if the mechanism had jammed," he points out, "that doesn't mean to say the rig's going to come crashing down on top of you, does it? It just means you're stuck there for a while. So what? You know, I've thought about this. They say that courage is a virtue, right?

"But what if you don't need courage? What then? What if you don't have fear to start with? If you don't have fear to start with, you don't need courage to overcome it, do you? The concrete and tire stunts wouldn't have bothered me, mate. They're just mind games. But that doesn't make me brave. If I couldn't give a shit in the first place, how can it?

"So you see, I just don't buy it. It seems to me that the reason you harp on about courage all the time, the reason people feel they need it, is to bring yourself up to the level I function at naturally. You may call it a virtue. But in my book, it's natural talent. Courage is just emotional blood doping."

Mindfulness

Sitting on a sofa opposite a six-foot-two psychopathic skinhead as he positions a sizable psychological magnet by the side of your moral compass isn't exactly comfortable. Of course, I'm well aware of the psychopath's powers of persuasion. But even so, I can't help thinking that Jamie has a point. What a "hero" might do against the muffled synaptic screaming of hardwired survival instincts, a psychopath might do in silence—without even breaking a sweat. And to set the compass spinning even faster, Leslie introduces another existential conundrum to the proceedings.

"But it's not just about functionality, though, is it?" he demurs. "The thing about fear, or the way I understand fear, I suppose—because, to be honest, I don't think I've ever really felt it—is that most of the time it's completely unwarranted anyway. What is it they say? Ninety-nine percent of the things people worry about never happen. So what's the point?

"I think the problem is that people spend so much time worrying about what might happen, what might go wrong, that they completely lose sight of the present. They completely overlook the fact that, actually, right now, everything's perfectly fine. You can see that quite clearly in your interrogation exercise. What was it that chap told you? It's not the violence that breaks you. It's the threat of it. So why not just stay in the moment?

"I mean, think about it. Like Jamie says, while you were lying under that lump of concrete—or rather, what you thought was concrete—nothing bad was really happening to you, was it? Okay, a four-poster might've been more relaxing. But actually, if you'd been asleep, you'd really have been none the wiser, would you?

"Instead, what freaked you out was your imagination. Your brain was on fast-forward mode, whizzing and whirring through all the possible disasters that might unfold. But didn't.

"So the trick, whenever possible, I propose, is to stop your brain from running on ahead of you. Keep doing that and, sooner or later, you'll kick the courage habit, too."

"Or you can always use your imagination to your advantage," interjects Danny. "Next time you're in a situation where you're scared, just think: 'Imagine I didn't feel this way. What would I do then?' And then just do it anyway."

Good advice—if you've got the balls to take it.

Listening to Jamie, Leslie, and Danny, you might well be forgiven for thinking you were in the presence of greatness: of three old Buddhists well on their way down the Eightfold Path to nirvana. Of course, they're anything but. Yet anchoring your thoughts unswervingly in the present, focusing exclusively, immediately, on the here and now, is a cognitive discipline that psychopathy and spiritual enlightenment have in common.

Mark Williams, professor of clinical psychology in the Department of Psychiatry at the University of Oxford, incorporates this principle of centering in his mindfulness-based cognitive-behavioral therapy (CBT) program for sufferers of anxiety and depression.

"Mindfulness," I tease Mark, in his office at the Warneford Hospital, "is basically Buddhism with a polished wooden floor, isn't it?"

He offers me a sweet roll.

"You forgot the spotlights and the plasma TV," he counters. "But yes, there's a certain whiff of the East in a lot of the theory and practice."

Mark gives me an example of how mindfulness-based CBT might help someone overcome a phobia, like a fear of flying, for instance. Jamie, Leslie and Danny couldn't have put it better.

"One approach," Mark explains, "might be to take the person on a plane and seat them next to a flying buff. You know, someone who absolutely loves being up in the air. Then, midflight, you hand them a pair of brain scans. One of them depicts a happy brain. The other one depicts an anxious brain. A brain in a state of terror.

" 'This pair of pictures,' you tell them, 'represents exactly what's going on in each of your heads right now, at this precise moment in time. So obviously, because they're so different, neither of them really means anything, do they? Neither of them predicts the physical state of the plane.

" 'That truth is in the engines.

" 'So, what *do* they signify?' you ask them. 'Well,' you explain, 'what, in fact, they do represent is . . . precisely what you're holding in your hands. A brain state. Nothing more. Nothing less. What you're feeling is simply just that. A feeling. A neural network, an electrical ensemble, a chemical configuration, caused by thoughts in your head that drift in and out, that come and go, like clouds.

"'Now, if you can bring yourself round to somehow accepting that fact; to dispassionately observe your inner virtual reality; to let the clouds float by, to let their shadows fall and linger where they please, and focus, instead, on what's going on around you—each pixelated second of each ambient sound and sensation—then eventually, over time, your condition should begin to improve.'"

Action

Jamie and the boys' pragmatic endorsement of the principles and practices of mindfulness—though not, necessarily, of the precise existential variety that a distinguished Oxford professor might extol—is typical of the psychopath. Their rapacious proclivity to live in the moment, to "give tomorrow the slip and take today on a joyride" (as Larry rather whimsically puts it), is well documented—and at times (therapeutic implications aside) can be stupendously beneficial.

Take the financial world, for instance. Don Novick was a trader for sixteen years, and didn't lose a penny in any of them. He is also, as it happens, a psychopath. These days—retired, though still only forty-six—he lives quietly in the Scottish Highlands, adding to his wine cellar and collecting vintage watches.

I call Don a psychopath because that's what he calls himself. At least, he did the first time I met him. So to be on the safe side, I decided to run a few tests. The results turned out to be positive.

Sitting in one of the drawing rooms of his secluded Jacobean castle—the driveway's so long it could do with a couple of service stations—I ask Don, quite literally, the million-dollar question. What, precisely, is it that makes a successful trader? I'm not too interested in the difference between good and bad, I point out. More in the difference between good and really good.

Though he doesn't name names, he has no hesitation in answering the question objectively, from a qualitative, analytical standpoint.

"I would say that one of the biggest differences when it comes to separating out the really good traders is how they seem at close of play, when trading has finished and they're turning it in for the day," he tells me. "You know, trading is a profession that, if you're the least bit vulnerable mentally, can completely undo you. I've seen traders crying and being physically sick at the end of a hard session. The pressure, the environment, the people . . . it's all pretty brutal.

"But what you find with the guys at the very top is that at the end of the day, when they're heading out the door, you just don't know. You can't tell by looking at them whether they've raked in a couple of billion or whether their entire portfolio has just gone down the tubes.

"And there it is in a nutshell. Therein lies the fundamental principle of being a good trader. When you're trading, you cannot allow any members of your brain's emotional executive committee to knock on the door of the decision-making boardroom, let alone take a seat at the table. Ruthlessly, remorselessly, relentlessly, you have to stay in the present. You can't let what happened yesterday affect what happens today. If you do, you'll go under in no time.

"If you're prone to emotional hangovers, you're not going to last two seconds on the trading floor."

Don's observations, coming, as they do, from sixteen years on the fiscal razor's edge, are strongly reminiscent of the lab-based results from Baba Shiv, Antoine Bechara, and George Loewenstein's "gambling game" study. Logically, of course, the right thing to do was to invest in every round. But as the game panned out, some of the participants began declining the opportunity to gamble, preferring instead to conserve their winnings. They began, in other words, to "live in the past"—allowing, in Don's words, members of their brain's emotional executive committee to knock on the door of the decision-making boardroom. Bad move.

But other participants continued to live in the present. And, at the conclusion of the study, boasted a pretty healthy profit margin. These "functional psychopaths," as Antoine Bechara referred to them—individuals who are either better at regulating their emotions than others or, alternatively, don't experience them to the same degree of intensity—continued to invest and treated each new round as if it were the first.

Oddly enough, they went from strength to strength. And, exactly as Don would have predicted (indeed, did predict when I told him about the experiment), wiped the floor with their cagier, more risk-averse rivals.

But the story doesn't end there. Several years ago, when news of this study first hit the popular press, it carried a headline that grabbed a few headlines itself: "Wanted: Psychopaths to Make a Killing in the Market." According to Don, the caption has hidden depths.

"A professional killer, like an executioner, for instance, probably has no feeling at all after taking someone's life," he explains. "Plausibly, remorse or regret just don't come into the equation. It's a similar story with traders. When a trader completes a trade, he'll call it an 'execution.' That's common trading parlance. And once a trade has been executed, the really good traders—the kind of guys that you're interested in—will have no compunction at all about getting out. About the whys and wherefores, the pros and cons—about whether it's right or wrong.

"And that's completely irrespective, going back to what I was saying earlier, of how that trade has gone—whether they've made a couple of billion or whether they've thrown it down the toilet. Exiting a trade will be a cool and clinical decision that has no subsequent emotion, no lingering psychological aftereffects attached to it whatsoever . . .

"I think the idea of killing professionally, be it in the market or elsewhere, demands a certain ability to compartmentalize. To focus on the job at hand. And, when that job is finished, to just walk away and forget it ever happened."

Of course, living in the past is just one side of the equation. Living in the future, getting "ahead of ourselves," allowing our imagination to run riot—as mine had done under that pallet of reinforced concrete, or whatever the hell it was—can be equally incapacitating. Studies, for instance, of cognitive and emotional focus in the context of dysfunctional decision making have shown that whenever we evaluate common, everyday behaviors—things like diving into a swimming pool, or picking up the phone and delivering bad news—the imagined, potential reality is significantly more discomfiting than the real one.

Which explains, of course, our unquenchable urge to procrastinate much of the time.

But psychopaths never procrastinate.

Just one of the reasons why, if you recall the words of Richard Blake from earlier, my host at Broadmoor and one of the clinical team in the Paddock Centre, they tend to excel at activities on the ward. Psychopaths need to do something. Nothing just isn't an option.

"Feeling good is an emergency for me," Danny had commented, as he'd slammed in his fourth goal for Chelsea. "I like to ride the roller coaster of life, spin the roulette wheel of fortune, to terminal possibility."

He frowned, and adjusted his baseball cap.

"Or at least I did"—he shrugged—"till I got in here."

Coming from a psychopath, it's not an untypical statement—one we could all perhaps do with taking on board just that little bit more in our lives.

"When I was a kid," Larry tells me, "we used to go on holiday every year to Hastings. One day—I'll never forget it—I watched my sister playing in the sea, and this big wave came in and hit her. She ran out crying, and that was that. She never went in again. When I saw what had happened—and I couldn't have been more than seven or eight at the time—I remember thinking to myself: 'If you stand where the waves break, you're going to get hurt. So you've got two choices. You can either stay on the shore and not go in at all. Or you can go out further so the waves lift you up and then crash and break behind you.'"

Jamie gets to his feet.

"The secret, of course, is not to go out too far," he grunts. "Otherwise you wash up in this place."

SOS Mentality

"Well, you know where I am. I ain't going nowhere."

Jamie and I are shaking hands. I've just told him I'll definitely look him up next time I'm passing, and he's filled me in on his movements. Larry and Leslie have already bowed out. Leslie quite literally, with a sweeping genuflection. Larry with a sturdy salute. Maybe the old boy was a former sea captain after all. Danny's returned to the soccer.

Back in the corridors and security-infested wormholes that connect the DSPD unit with the outside world, I feel a bit like a spaceman on reentry.

"Settle in okay?" Richard inquires as we jangle our way back to clinical psychology suburbia.

I smile. "Started to feel quite at home."

As the train picks up speed for London, I study the expressions of those sitting around me: commuters, mostly, on their way home from work. Some are tense and anxious. Others are tired and drawn. You don't see many of those kinds of faces at Psychopath School.

I fire up the laptop and punch in some thoughts. An hour or so later, as we pull into the station, I have the template for what I call an "SOS" mentality: the psychological skill set to Strive, Overcome, and Succeed.

I label the skill set the Seven Deadly Wins—seven core principles of psychopathy that, apportioned judiciously and applied with due care and attention, can help us get exactly what we want; can help us respond, rather than react, to the challenges of modern-day living; can transform our outlook from victim to victor, but without turning us into a villain:

1. Ruthlessness
2. Charm
3. Focus
4. Mental toughness

5. Fearlessness
6. Mindfulness
7. Action

Without a shadow of a doubt, the power of the skill set lay squarely in its application. Certain situations would inevitably call for more of some traits than others, while within those sets of circumstance, some sub-situations, going back to our trusty mixing desk analogy, would plausibly demand higher or lower output levels of whichever traits were selected. Cranking up the ruthlessness, mental toughness, and action dials, for instance, might make you more assertive—might earn you more respect among your work colleagues. But ratchet them up too high and you risk morphing into a tyrant.

Then, of course, there was the opposite consideration of being able to turn them back down—of fading in and out and appropriately contouring the soundtrack. If the lawyer, for example, whom we met in chapter 4 was as ruthless and fearless in everyday life as he evidently was in the courtroom, he would've soon ended up needing a lawyer of his own. The secret, unquestionably, was context. It wasn't about being a psychopath. It was rather about being a method psychopath. About being able to step into character when the situation demanded it. But then, when the exigency had passed, to revert to one's normal persona.

That, of course, was where Jamie and the boys had gone wrong. Rather than having trouble twiddling the faders up higher, theirs, in contrast, were permanently stuck on max: a manufacturing error with decidedly unfortunate consequences. As Jamie had articulated when I'd first arrived in Broadmoor, the problem with psychopaths isn't that they're chock-full of evil. Ironically, it's precisely the opposite: they have too much of a good thing.

The car is to die for. It's just too fast for the road.

SEVEN

SUPERSANITY

> Life should not be a journey to the grave with the intention of arriving safely in a pretty and well preserved body, but rather to skid in broadside in a cloud of smoke, thoroughly used up, totally worn out, and loudly proclaiming "Wow! What a Ride!" —HUNTER S. THOMPSON

Generation P

At the back of the chapel of Magdalen College, Oxford, there sits a prayer board. One day, amid the numerous petitions for divine intervention, I noticed this one: "Lord, please make my lottery numbers come up, then you won't ever have to hear from me again." Strangely enough, it was the only one that God had replied to. Here is what he wrote: "My son, I like your style. In this wretched, mixed-up world, which causes me so much sorrow, you've put a smile on my face. Hell, I WANT to hear from you again. So better luck next time you cheeky bastard! Love God."

Anyone who thought that God didn't have a sense of humor might want to think again. And anyone who thought that God was so remote from the world that he didn't take a personal interest in the trifling concerns of his lost, lamentable little children might also wish to reconsider. Here, quite clearly, the Almighty sees fit to present a different side of Himself: as a shrewd, tough, no-nonsense operator able to give as good as He gets and with more than a passing knowledge of human psychology. If that sounds like the kind of God who's not afraid to twiddle those dials on the mixing desk up and down when the situation calls for it, then you're not mistaken.

In 1972, the writer Alan Harrington published a little-known book called *Psychopaths.* In it, he advanced a radical new theory of human evolution. Psychopaths, contended Harrington, constitute a dangerous new breed of *Homo sapiens*: a made-to-measure Darwinian contingency plan for the cold, hard exigencies of modern-day survival. An indomitable Generation P.

Key to his thesis was the progressive, insidious weakening, as he saw it, of the primeval ionic bonds—ethical, emotional, existential—that for century upon century, millennium upon millennium, had bound humanity together. Once, argued Harrington, when Western civilization subscribed to the traditional bourgeois mores of hard work and virtue seeking, the psychopath was confined to the margins of mainstream society. He was condemned, by his fellow right-minded citizens, as either a madman or an outlaw. But as the twentieth century unfolded and society, over time, became ever more fast and loose, psychopaths came in from the cold.

For a Cold War novelist from a nonscientific background, Alan Harrington certainly knew his stuff. His depiction of the psychopath rivals—in fact occasionally even surpasses, given its diverse, eclectic brushstrokes—many of the portraits one sometimes reads today. The psychopath, as Harrington defines him, is the "new man": a psychological superhero free from the shackles of anxiety and remorse. He is brutal, bored, and adventurous. But also, when the situation demands it, beatific.

Harrington cites some examples: "Drunkards and forgers, addicts, flower children . . . Mafia loan shark battering his victim, charming actor, murderer, nomadic guitarist, hustling politician, the saint who lies down in front of tractors, the icily dominating Nobel Prize winner stealing credit from laboratory assistants . . . all, all doing their thing."

And all, all, without the slightest care in the world.

Saint Paul—The Patron Saint of Psychopaths

Harrington's inclusion of saints on the list is no accident. It's also no one-off. Throughout his book, the cool, iridescent prose is littered with comparisons between psychopaths and the spiritually enlightened. Not all of them his own.

He quotes, for instance, the physician Hervey Cleckley, whom we met in chapter 2, compiler, in his 1941 classic *The Mask of Sanity*, of one of the first clinical descriptions of psychopathy:

"What he [the psychopath] believes he needs to protest against turns out to be no small group, no particular institution, or set of ideologies, but human life itself. In it he seems to find nothing deeply meaningful or persistently stimulating, but only some transient and relatively petty pleasant caprices, a terribly repetitious series of minor frustrations, and ennui . . . Like many teenagers, *saints* [author's emphasis], history-making statesmen, and other notable leaders or geniuses, he shows unrest: he wants to do something about the situation."

Harrington also quotes Norman Mailer: "[The psychopath] is an elite with the potential ruthlessness of an elite . . . His inner experience of the possibilities within death is his logic. So, too, for the existentialist . . . And the *saint* and the bullfighter and the lover."

The implications are intriguing. Is it possible, Harrington asks, that the saint and the psychopath somehow constitute two transcendental sides of the same existentialist coin? Is it possible "whether we want to admit it or not, for the most wicked, wholly inexcusable psychopath to murder his way into a state of grace? Achieve a sort of purity by terrible means? Be transformed by his ordeal, and the ordeals he imposed on others, into a different person, his spirit cleansed by theater, publicity, fame, terror?"

Though contrary, perhaps, to their delicate intellectual sensibilities, New Testament scholars might struggle to disagree. Two thousand years ago, a certain Saul of Tarsus sanctioned the deaths of countless numbers of Christians following the public execution of their leader—and could today, under the dictates of the Geneva Convention, have been indicted on charges of genocide.

We all know what happened to him. A dazzling conversion as he journeyed on the road to Damascus* transformed him, quite literally overnight, from a murderous, remorseless tentmaker into one of the most important figures in the history of the Western world. Saint Paul, as he's more commonly referred to today, is the author of just over half of the entire New Testament (fourteen of the twenty-seven books that comprise the corpus are attributed to him); is the hero of another, the Acts of the Apostles; and is the subject of some of the best stained glass in the business.

But he was, in addition, almost certainly a psychopath. Ruthless, fearless, driven, and charismatic, in equal measure.

Let's take a look at the evidence. Paul's apparent predilection, both on the open road and within seething inner cities, for dangerous, inhospitable areas put him at constant risk of random, violent assault. Add to that the fact that he was shipwrecked a grand total of three times during his travels around the Mediterranean basin, on one occasion spending twenty-four hours adrift in the open sea before being rescued, and a picture begins to emerge of a man with little or no concern for his own safety.

There's the habitual lawbreaker who seems incapable of learning the error of his ways (either that or he just didn't care). Paul was imprisoned multiple times during his ministry, spending, in total, an estimated six years behind bars; was brutally flogged (five times receiving the maximum thirty-nine lashes: too many might kill a person); was, on three occasions, beaten with rods. And was once, in the city of Lystra, in what is now modern Turkey, stoned by a mob so viciously that he was, by the time they were finished, given up for dead and dragged outside the city, as was the custom.

* Modern-day experts in the field of neurotheology consider Saul's experience to be more symptomatic of the onset of temporal lobe epilepsy than of any genuine encounter with the Divine. The "light from Heaven," the auditory hallucinations ("Saul, Saul, why persecutest thou Me?"), and his subsequent temporary blindness are certainly compatible with such a diagnosis—as is Saul's own mysterious, health-related allusion (2 Corinthians. 12:7–10) to a "thorn in the flesh," a "messenger of Satan," "to keep me from becoming conceited."

Scripture records what happened next: "But when the disciples gathered about him, he rose up and entered the city, and on the next day he went on with Barnabas to Derbe" (Acts 14:20).

Would you calmly reenter a city whose inhabitants had just tried to stone you to death? Not sure I would.

And we're not finished yet. There's the peripatetic drifter who was continually on the move due to threats against his life. When the governor of Damascus placed a cordon around the city to arrest him, he made his escape in a basket through a gap in the city walls.

There's the cold, calculating, political mover and shaker, unafraid to tread on the feelings and sensibilities of others, no matter how important or personally loyal they were. Paul's bust-up with Saint Peter in Antioch, in which he accused Peter to his face of being a hypocrite in forcing Gentiles to adopt Jewish customs when he himself lived like a Gentile, is described by L. Michael White, professor of classics and religious studies at the University of Texas at Austin, in his book *From Jesus to Christianity* as "a total failure of political bravado, and Paul soon left Antioch as persona non grata, never again to return."

Finally, there's the remorseless, unblinking maneuvering of the shadowy psychological cat burglar. The silky-smooth self-presentation skills of the expert manipulator.

Recall the words of master con man Greg Morant? One of the most powerful weapons in the grifter's unholy arsenal is a good "vulnerability radar." They could just as easily have been Paul's.

Or, to put it another way:

"To the Jews I became as a Jew, in order to win Jews. To those under the law I became as one under the law (though not being myself under the law) that I might win those under the law. To those outside the law I became as one outside the law (not being outside the law of God but under the law of Christ) that I might win those outside the law. To the weak I became weak, that I might win the weak. I have become all things to all people" (1 Corinthians 9:20–22).

If it really was Jesus on that road to Damascus and he wanted an emissary to help him spread the word, he couldn't have picked a better man for the job. Nor, among Christians, a more feared or unpopular one. At the time of his conversion Paul, without doubt, was at the height of his persecutory powers. In fact, the very reason he was going to Damascus in the first place was to instigate more bloodshed. Coincidence that his ministry started there?

Not all psychopaths are saints. And not all saints are psychopaths. But there's evidence to suggest that deep within the corridors of the brain, psychopathy and sainthood share secret neural office space. And that some psychopathic attributes—stoicism, the ability to regulate emotion, to live in the moment, to enter altered states of awareness, to be heroic, fearless, yes, even empathic—are also inherently spiritual in nature, and not only improve one's own well-being, but also that of others.

If you need any convincing, you've just got to look at the Magdalen College prayer board once in a while.

Seeing Red Swings It for the Champion

The ability to smile in the face of adversity has long been regarded as a measure of spiritual intelligence. Take, for instance, the words of the poet Rudyard Kipling, the last thing you see as a player before stepping out onto center court at Wimbledon:

> *"If you can meet with Triumph and Disaster*
> *And treat those two impostors just the same . . ."*

But though such a mind-set is usually associated with saints, the link with psychopaths is somewhat less often conjectured.

In 2006, Derek Mitchell at University College, London, decided to buck the trend—and presented two groups of participants, psychopaths and non-psychopaths, with a procedure known as the Emotional Interrupt Task (EIT). The EIT is a reaction time test of discriminatory ability. Typically, volunteers sit in front of a computer screen and press keys with either their left or their right index fingers depending on whether a particular kind of shape, usually a circle or a square, flashes up in front of them.

Pretty simple, you might think. But in actual fact, the task can be rather tricky.

The reason for this is that the shapes don't appear on their own. Instead, each circle or square is sandwiched, for a couple of hundred milliseconds at a time, between different pairs of images—usually faces. Either two positive images (smiling faces), two negative images (angry faces), or two neutral images (expressionless faces), respectively.

Most people find the emotional images a problem. Simply because of *that*: they're emotional—and distracting. But if, Mitchell hypothesized, psychopaths really are as unfazed and laid back as their reputations suggest, really can take the rough with the smooth, this shouldn't, in their case, be true. They should, in fact, respond faster and more accurately, compared to controls—that is, they should be less distractible—on those trials where the circle or square is flanked by either two positive or two negative images. Images, in other words, that, one way or another, conceal an emotional valence. In contrast, Mitchell suggested, this difference between psychopaths and non-psychopaths should disappear on the neutral trials, where distraction is less of an issue.

As it turned out, this is exactly what he found. Whenever the circle or square was flanked by an emotionally charged image, the psychopaths, exactly as predicted, were better at differentiating the targets than the non-psychopaths—and quicker, too. They were, as Kipling might've put it, better at keeping their heads while others were losing theirs.

Stoicism is a quality greatly prized by society. And with good reason. It can come in handy in all sorts of ways: during bereavement, after a breakup, at the poker table—even, at times, when you're writing a book. But as a long-suffering follower of the England soccer team, and a veteran of more penalty shoot-out debacles than I care to remember, it's the relationship between stoicism and sport that's perhaps most salient to me.

And I'm not just talking from a spectator's point of view. As a psychological prism, sport is second to none in dispersing stoicism into its two constituent wavelengths, fearlessness and focus, which themselves comprise elements of both psychopathy and spiritual acumen.

"Do you not know that in a race all the runners run, but only one gets the prize?" wrote Saint Paul. "Run in such a way as to get the prize . . . I am like a boxer who does not waste his punches. I harden my body with blows and bring it under complete control" (1 Corinthians 9:24, 26). That Kipling's words hang above center court is certainly no coincidence . . . nor are they exclusive to tennis.

"Play like it means nothing, when it means everything," replied the snooker legend Steve Davis when he was asked to spill the beans on sporting greatness. "Let go" of bad shots—and, for that matter, good ones—and focus your attention 100 percent on the next one.

The same is true in golf.

In 2010, the South African Louis Oosthuizen was a rank outsider to win the British Open Championships at St. Andrews. After a string of disappointments in the events leading up to the tournament, he was fully expected, even with a four-shot lead, to crack under the pressure of the cutthroat final round. But he didn't. And the reason for that was surprisingly, though deceptively, simple: a small red spot, conspicuously located just below the base of his thumb, on his glove.

The idea of the spot came from Karl Morris, a Manchester-based sports psychologist, who was called in by Oosthuizen to help him touch base with what one might reasonably describe as his hidden inner psychopath. The goal was to center his mind on playing the shot at hand, rather than obsessing, at exactly the wrong moment, about the consequences. So Morris devised a plan. Whenever Oosthuizen was about to take a swing, he was to focus his attention, calmly and steadily, on the dot. The dot was all that mattered at that point in time. He wasn't to play the shot. The shot, instead, was to play him.

He won by seven strokes.

Oosthuizen's red spot is a classic example of what's known in sports psychology as a process goal—a technique by which the athlete is required to focus on something, however minor, to prevent them from thinking about other things: in Oosthuizen's case, all the ways he could possibly screw up the shot. It anchors the athlete firmly in the here and now—before the shot is actually played, before the move is actually made, and, most important of all, before confidence begins to fade. In fact, this ability to concentrate purely on the task at hand—what the Hungarian psychologist Mihály Csíkszentmihályi calls "optimal experience," or "flow"—is one of a number of key techniques that performance psychologists now work on, not just in golf, but among high-level competitors in all areas of sport.

In moments of flow, the past and the future evaporate as abstractions. All that remains is an intense, uncanny, attention-devouring present, an overwhelming feeling of being "in the zone." This is the union, the enchanted consummation, of mind, body, and game: what's known in the trade as the "Golden Triangle" of performance, a trance-like state of effortless action and reaction, where time and self converge—and one is in control, but not in control, at the same time.

It has a telltale neural signature in the brain.

In 2011, Martin Klasen at Aachen University discovered that moments of flow possess a unique physiological profile. Using fMRI to observe the brains of video game players in action, he found that periods of heightened focus and concentration are accompanied by a reduction of activity in the anterior cingulate cortex—the brain's error-detection and conflict-monitoring hardware—indicative of an increase in attention and the suppression of distracting and non-task-relevant information.

But that's not all. A similar pattern has also been found in the brains of criminal psychopaths.

In the same year that Klasen was playing video games, Kent Kiehl dusted off that eighteen-wheel fMRI juggernaut of his and hit the roads of New Mexico armed with a new experiment. Kiehl was interested in what, precisely, makes psychopaths tick when it comes to moral decision making. Are they really ice-cold under pressure? Are they really better at coming up with the goods when the chips are down and time is of the essence? If so, why? Could it possibly be something that's hardwired into their brains? A triumph of cold-blooded cognitive reasoning over warm-blooded emotional processing?

To find out, he presented psychopaths and non-psychopaths with two different types of moral dilemmas: what he termed "high-conflict (personal)" and "low-conflict (personal)" dilemmas, respectively, examples of both of which follow.*

High-Conflict (Personal)

Enemy soldiers have taken over your village. They have orders to kill everyone they find. You and some others are hiding in a basement. You hear the soldiers enter the house above you. Your baby begins to cry loudly. You cover his mouth to block the sound. If you remove your hand from his mouth he will cry loudly and the soldiers will hear. If they hear the baby they will find you and kill everyone, including you and your baby. To save yourself and the others, you must smother your baby to death.

Is it morally acceptable for you to smother your child in order to save yourself and the other people?

Low-Conflict (Personal)

You are visiting your grandmother for the weekend. Usually she gives you a gift of a few dollars when you arrive, but this time she doesn't. You ask her why not and she says something about how you don't write her as many letters as you used to. You get angry and decide to play a trick on her.

You take some pills from the medicine cabinet and put them in your grandmother's teapot, thinking that this will make her very sick.

Is it morally acceptable for you to put pills in your grandmother's teapot in order to play a trick on her?

* Kiehl and his coauthors also included a third type of dilemma, which they termed "impersonal." This took the form of the original version of the Trolley Problem devised by Philippa Foot (see chapter 1), in which the choice (initiated by the flick of a switch) is whether to divert a runaway train away from its present course of killing five people onto an alternate course of killing just one.

The prediction was simple. If psychopaths were less fazed by the extraneous emotional exigencies of the moment and had the cold, hard edge over the rest of us when it came to life-and-death decision making, then the most marked difference between their performance and that of the non-psychopaths, Kiehl surmised, should manifest itself when it came to the high-conflict (personal) dilemmas—when the heat is turned up to max and the problem is closest to home. This, it turned out, is exactly what he found (see figure 6.1).

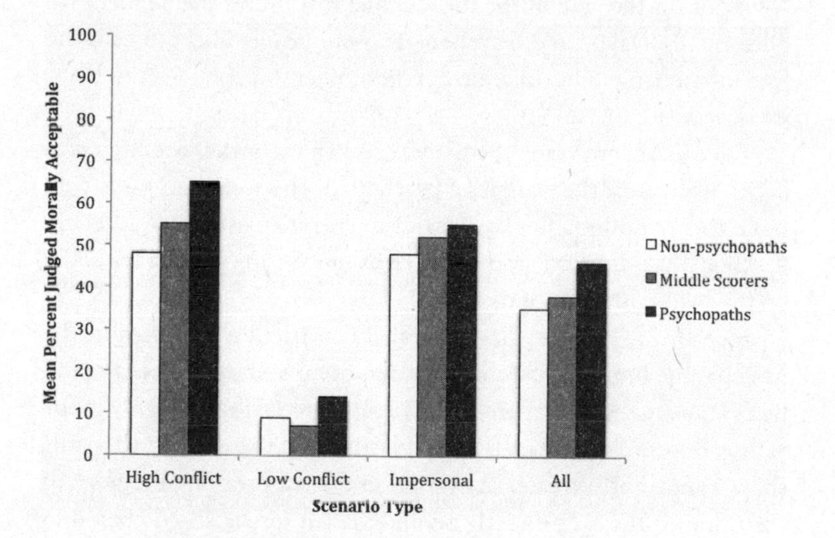

Figure 6.1. Psychopaths are less morally squeamish—but only when playing for high stakes. (Adapted from Ermer et al., 2011)

In the high-conflict scenarios, the psychopaths did, indeed, rate a significantly greater number of utilitarian judgments as "morally acceptable" than the non-psychopaths. They were better at smothering babies, or at least dealing with the pain of such an action, than their more ethically squeamish counterparts. And presumably they would be better at staying alive, and preserving the lives of their fellow basement stowaways, were the scenarios to play out for real.

But there was more. Just as I had found with the *William Brown* example in chapter 3, Kiehl and his coworkers also discovered that the psychopaths, as well as having fewer moral hygiene issues in general than the non-psychopaths, took considerably less time to evaluate the conundrums put in front of them. They were quicker at reaching a decision as to the appropriate course of action. Not only that, but such attenuated response times were accompanied, just as Martin Klasen had found under conditions of flow, by reduced activity in the anterior cingulate cortex.

But—and here's the rub—only when it came to the high-conflict scenarios. In the case of the low-conflict dilemmas, the deliberative differential disappeared. Psychopaths were just as likely to veto the idea of putting pills in your grandmother's teapot as the non-psychopaths.

The conclusion seems pretty clear. When the stakes are high and backs are against the wall, it's a psychopath you want alongside you. But if there's nothing to play for and you're on an even keel, forget it. Psychopaths switch off, and take just as much time getting the show on the road as the rest of us.

Indeed, EEG studies have revealed consistent differences in the way that the brains of psychopaths and non-psychopaths respond to tasks and situations that are either highly interesting or highly motivating, respectively. When the handwriting's on the wall, psychopaths show significantly greater activation in the left prefrontal regions of their brains (the area directly behind the left forehead) compared to non-psychopaths—cerebral asymmetry associated with considerable reduction in anxiety, enhanced positive affect, increased focusing of attention, and orientation to reward. And also, it would seem, with elevated spiritual states. The neuroscientist Richard Davidson, at the University of Wisconsin, has discovered precisely the same profile in elite Buddhist monks, the spiritual Olympians of the High Himalayas, while they're immersed in deep meditation.

"There is a lot of evidence [to suggest] that the best sportsmen and women have [developed] psychological skills that allow them to concentrate and to control anxiety," explains Tim Rees, a sports psychologist at the University of Exeter. Moreover, he adds, "there is also a lot of evidence to show that, once someone gets to a certain level of skill, it is the differences in their psychological approach that differentiates people at the very top."

The mind-set that separates the great from the good, and as Kent Kiehl showed us, in some critical situations, the living from the dead, is inherently psychopathic in nature.

And also inherently spiritual.

Stop All the Clocks

The link made by Csíkszentmihályi and others between "staying in the present" and the absence of anxiety is, of course, hardly new. The practice of "right mindfulness," for example, constitutes the seventh step of the Noble Eightfold Path, one of the principal teachings of Siddhartha Gautama, the Buddha, some two and a half thousand years ago.

In his book *The Noble Eightfold Path: The Way to the End of Suffering*, Bhikkhu Bodhi, a monk of the Theravada tradition, describes what the practice entails:

"The mind is deliberately kept at the level of *bare attention*, a detached observation of what is happening within us and around us in the present moment. In the practice of right mindfulness the mind is trained to remain in the present, open, quiet, and alert, contemplating the present event. All judgments and interpretations have to be suspended, or if they occur, just registered and dropped."

According to the Mahāsatipaṭṭhāna Sutta, one of the core discourses in the Pāli canon of Theravada Buddhism, such training, consistently applied, eventually leads to "the arising of insight and the qualities of dispassion, non-clinging, and release."

Qualities, as we've seen, that psychopaths appear to possess naturally.

But the similarities between the Western, psychopathic mentality and the transcendental mind-sets of the East don't end there. More recently, psychologists such as Mark Williams at the University of

Oxford—who, if you recall, we met in the previous chapter—and the aforementioned Richard Davidson have begun the innovative, integrative, though empirically demanding process of harnessing the restorative properties of Buddhist meditation practices within a more systematized, therapeutic, and clinically oriented framework.

So far it seems to be working.

Mindfulness-based intervention, as we saw a little earlier, has been shown to be a particularly effective metacognitive strategy when dealing with the symptoms of anxiety and depression . . . two conditions that psychopaths are singularly immune to.

The fundamental principles of the therapy, as one would expect, are heavily derivative of the traditional Buddhist teachings already outlined. But there's an added ingredient, a kind of naive, childlike inquisitiveness, which is strongly reminiscent of the core "openness to experience" factor of the Big Five personality structure that we explored in chapter 2. And which psychopaths, if you recall, score very high on.

"The first component [of mindfulness] involves the self-regulation of attention so that it is maintained on immediate experience," explains psychiatrist Scott Bishop in one of the seminal papers on the subject back in 2004, "thereby allowing for increased recognition of mental events in the present moment. The second component involves adopting a particular orientation toward one's experiences in the present moment, an orientation that is characterized by curiosity, openness, and acceptance."

Or, as the Zen Buddhist masters of the martial arts would say, *shoshin*: "beginner's mind."

"In the beginner's mind there are many possibilities," elucidated Shunryu Suzuki, one of the most celebrated Buddhist teachers of recent times. "In the expert's mind there are few."

And few would disagree. When Dickens decided to send Scrooge the ghosts of the past, present, and future, he chose the three specters that haunt all of us. But anchor your thoughts entirely in the present, screen out the chatter of the querulous, recriminative past and the elusive, importunate future, and anxiety begins to subside. Perception begins to sharpen. And the question becomes one of utility: what we do with this 'now', this enormous, emphatic present, once we have it. Do we "savor" the moment like a saint? Or "seize" it like a psychopath? Do we reflect on the nature of experience? Or do we focus our attention entirely on ourselves in the frenetic pursuit of instant gratification?

Several years ago, as I documented in *Split-Second Persuasion*, I traveled to a remote monastery in Japan in search of the answer to a mystery. The mystery in question concerned a test: a test undertaken by those on the rarefied spiritual ice fields of the higher martial arts.

The test involves one man kneeling down—arms by his side, blindfolded—while another man stands behind him with a samurai sword raised directly above his head. At a moment of his choosing, unbeknownst to his vulnerable adversary, the man standing behind will unleash the sword onto the kneeling man's frame, causing injury, probably death. Unless, that is, the blow is somehow deflected. And the swordsman then disarmed.

Such an undertaking appears impossible. And yet it isn't. The test I have just described is real: an ancient, exquisitely choreographed ritual, carried out in secret, unfathomable dojos in Japan and the High Himalayas, that those approaching greatness—those spectral mind whisperers miles above the black belt—routinely undergo.

These days, mercifully, the sword is made of plastic. But there was a time, long before the days of health and safety, when it was the real thing.

A shadowy sensei, well into his eighties, revealed the secret:

"One must empty one's mind totally," he told me, as we sat cross-legged in a garden of clouds and lilac, deep in the ancient beech forests of the Tanzawa Mountains. "One must focus purely on the now. When one enters a state like that, one is able to smell time. To feel its waves washing over one's senses. The tiniest ripple may be detected over great distances. And the signal intercepted. Often it appears that the two combatants move simultaneously. But this is not so. It is not difficult. With practice it can be mastered."

Reading back over what the ancient sensei told me, I'm strongly reminded of the words of the psychopathic neurosurgeon we encountered in chapter 4. Of course, I hadn't met him by the time I went to

Japan. But if I had, I would've recounted his description of how he sometimes feels before a difficult operation with great alacrity to my host. And the old man, in his monastic black *hakama* and blood-red kimono, would've smiled. The surgeon's account of the mental state he calls "supersanity"—an "altered state of consciousness that feeds on precision and clarity"—appears very similar to the frame of mind that the sensei was talking about: the mind state that must be entered into by the kneeling, blindfolded time-sommelier to disarm his sword-wielding assailant.

One is also reminded of the work of Joe Newman, who, if you recall, in his lab at the University of Wisconsin, has shown that it isn't so much the case that psychopaths don't *feel* anxiety in certain situations, but rather that they just don't notice the threat. Their attention is focused purely on the task at hand, and extraneous distractors are ruthlessly filtered out.

Usually, of course, when it comes to psychopaths, such focus is construed as malevolent: the glacial, guiltless killer cruising the city limits like some gas-guzzling praying mantis in search of the perfect victim; the genocidal dictator, oblivious to the rule of moral or civil law, hell-bent on silencing all dissenting opinion in his indomitable quest for cultural and political omnipotence.

Compassionate, transcendental, or spiritual connotations are rarely, if ever, considered.

But a number of studies have recently begun to shed new light on such a resoundingly unlikely possibility—and have begun to precipitate a gradual yet fundamental reevaluation of precisely what it means to be a psychopath.

Heroes and Villains

Mem Mahmut, at Macquarie University in Sydney, has turned up something extraordinary. Psychopaths, it appears, far from being callous and unemotional all the time, can actually, in the right kind of context, be more altruistic than the rest of us.

Mahmut conducted a study comprising a series of real-life scenarios in which people asked for help from passersby (unsuspecting volunteers who'd previously been tested for psychopathy and classified as high or low scorers).

But there was a catch. The people asking for help, just like the passersby who responded to them, weren't exactly random. They were, in fact, Mahmut's evil co-conspirators in a unique, fiendishly constructed experiment specifically designed to investigate the relationship between psychopathy and helping behavior.

The experiment consisted of three parts. In the first part, Mahmut's accomplices solicited help directly from the passersby by pretending to be lost and approaching to ask for directions. In the second part, the "request" for help was much less direct and explicit: a hapless female who'd dropped a bunch of papers. In the third part, the request was less explicit still: a lab researcher who'd supposedly broken her arm pretending to have difficulty with a variety of simple tasks—opening a water bottle, and entering the participant's name in a fake logbook, for instance—but bravely persevering in spite of her conspicuous injury.

Who, in these three different scenarios, Mahmut wanted to know, would be most likely to offer help: the remorseless, cold-hearted psychopaths or their warmer, more empathic counterparts?

The results of the study blew Mahmut away. In fact, so far off beam were they that he's still trying to get his head around them.

In the first part of the experiment, in which the accomplice asked for directions, the psychopaths, as predicted, offered less help than the non-psychopaths. No surprises there. In the second part, however—dropped papers—the altruism gap mysteriously disappeared. The psychopaths and the non-psychopaths offered an equal amount of help.

But it was in the third part, in which the accomplice feigned an injury, that the wheels really came off Mahmut's preconceived hypothesis that the psychopaths would be less obliging. In fact, the opposite turned out to be true.

The psychopaths showed a greater readiness to put themselves forward for water-bottle duty and to enter their own names into the logbook than the non-psychopaths. When the person asking for help was at their most vulnerable, and yet at the same time didn't endeavor to proactively solicit aid, the psychopaths came up with the goods. When it really mattered, they were significantly more likely to step up to the plate than were their (supposedly, at least) warmer, more empathic counterparts.

Unsurprisingly, perhaps, the results of Mahmut's experiment have certainly raised some eyebrows. One interpretation, of course, is that, as some enlightened (and no doubt rather embittered) soul once pointed out, there is no such thing as a truly altruistic act. There is always, however well camouflaged in the darker recesses of our dense psychological undergrowth, an ulterior self-serving, distinctly less honorable motive—and the psychopaths in Mahmut's study, with their finely tuned, highly sensitive long-range vulnerability antennae (recall the experiment conducted by Angela Book in which psychopaths were better able than non-psychopaths to pick out the victims of a violent assault merely from the way they walked), quite simply "smelled blood."

"It is pleasure that lurks in the practice of every one of your virtues," wrote the novelist W. Somerset Maugham in *Of Human Bondage*. "Man performs actions because they are good for him, and when they are good for other people as well they are thought virtuous . . . It is for your private pleasure that you give twopence to a beggar as much as it is for my private pleasure that I drink another whiskey and soda. I, less of a humbug than you, neither applaud myself for my pleasure nor demand your admiration."

On the other hand, however, there's evidence to suggest that Mahmut's incendiary findings are no flash in the pan. And that they mark the beginning of a welcome new shift in both empirical and theoretical focus: away from the conventionally pejorative physiological profiles cranked out by the neuroimaging brigade, toward a more applied, pragmatic research drive into functional "positive psychopathy." As a case in point, Diana Falkenbach and Maria Tsoukalas, at the John Jay College of Criminal Justice, City University of New York, have recently begun studying the incidence of so-called adaptive psychopathic characteristics in what they term "hero populations": in front-line professions such as law enforcement, the military, and the rescue services, for example.

What they've discovered jells nicely with the data that Mahmut's research has uncovered. On the one hand, though exemplifying a prosocial lifestyle, hero populations are tough. Perhaps unsurprisingly, given the level of trauma and risk such occupations entail, they show a greater preponderance of psychopathic traits associated with the Fearless Dominance and Coldheartedness subscales of the PPI (e.g., low anxiety, social dominance, and stress immunity), compared with the general population at large. These dials are turned up higher. On the other hand, however, they part company with criminal psychopaths in their relative absence of traits related to the Self-Centered Impulsivity subscale (e.g., Machiavellianism, narcissism, carefree nonplanfulness, and antisocial behavior). These dials are turned down lower.

Such a profile is consistent with the anatomy of the hero as portrayed by the psychologist Philip Zimbardo, founder of the Heroic Imagination Project—an initiative aimed at educating folk in the insidious techniques of social influence. Or more specifically, how to resist them. In 1971, in an experiment that has long since been inaugurated into psychology's hall of fame, Zimbardo constructed a simulated prison in the basement of Stanford University's psychology building and randomly assigned twelve student volunteers to play the role of prisoner, while another twelve were to play the role of guard.

After just six days, the study was abandoned. A number of the "guards" had begun to abuse the "prisoners," misusing their power simply because they had it. Forty years on, post–Abu Ghraib and the painful lessons learned, Zimbardo is engaged in a radically different project: to develop the "hero muscle" within all of us. Having uncorked the genie of the villain and the victim within, he now seeks to do the opposite: to empower everyday folk to stand up and make a difference when they might otherwise be silenced by fear. And not just when it comes to physical confrontations, but psychological confrontations, too—which, depending on the circumstances, can pose just as much of a challenge.

"The decision to act heroically is a choice that many of us will be called upon to make at some point in our lives," Zimbardo tells me. "It means not being afraid of what others might think. It means not being afraid of the fallout for ourselves. It means not being afraid of putting our necks on the line. The question is: Are we going to make that decision?"

Over coffee in his office, we talk about fear, conformity, and the ethical imperative of braving psychological as well as physical confrontation. Not unexpectedly, our old friend groupthink rears its head again, which occurs, as we saw in chapter 3 with the *Challenger* disaster, when the warped in-group forces of social gravitation exert such pressures on a collective as to precipitate—in the words of Irving Janis, the psychologist who carried out much of the early work on the process—"a deterioration in mental efficiency, reality testing, and moral judgment."

Zimbardo cites, as another case in point, the attack on Pearl Harbor by Japanese forces during the Second World War.

On December 7, 1941, the Imperial Japanese Navy launched a surprise attack against a United States naval base on the Hawaiian island of Oahu. The offensive was intended as a preemptive strike, aimed at precluding the U.S. Pacific Fleet from compromising planned Japanese incursions against the Allies in Malaya and the Dutch East Indies. It proved devastating. A total of 188 U.S. aircraft were destroyed; 2,402 Americans were killed and 1,282 injured—prompting then president Franklin D. Roosevelt to call the following day for a formal declaration of war on the Empire of Japan. Congress gave him the go-ahead. It took them less than an hour.

But might the attack on Pearl Harbor have been prevented? The catastrophic carnage and chaotic, combative consequences averted? There's evidence to suggest that it could, and that a constellation of groupthink factors—false assumptions, unchecked consensus, unchallenged reasoning biases, illusions of invulnerability—all contributed to a singular lack of precaution taken by U.S. Navy officers stationed in Hawaii.

In intercepting Japanese communications, for example, the United

States had reliable information that Japan was in the process of gearing itself up for an offensive. Washington took action by relaying this intelligence to the military high command at Pearl Harbor. Yet the warnings were casually ignored. The developments were dismissed as saber-rattling: Japan was merely taking measures to forestall the annexation of their embassies in enemy territories. Rationalizations included "The Japanese would never dare attempt a full-scale surprise assault against Hawaii because they would realize that it would precipitate an all-out war, which the United States would surely win"; and "Even if the Japanese were foolhardy enough to send their carriers to attack us [the United States], we could certainly detect and destroy them in plenty of time." History attests they were wrong.

As an example of the expediency of psychological troubleshooting, and of the spiritual qualities of fearlessness and mental toughness inherent in heroic action, both the *Challenger* and Pearl Harbor fiascoes provide intriguing parallels between the work of Philip Zimbardo and that of Diana Falkenbach and Maria Tsoukalas, mentioned earlier. Previously, in chapter 3, we explored the possibility that psychopathic characteristics such as charm, low anxiety, and stress immunity—the characteristics that Falkenbach and Tsoukalas identified in comparatively greater number in hero populations—may well, somewhat ironically, have managed to gain a toehold in our evolutionary gene pool through their propensity to facilitate conflict resolution. Dominant individuals among chimpanzees, stump-tailed monkeys, and gorillas all compete for mates, if you recall, by intervening in disputes among subordinates.

Yet an alternative explanation—and the two, of course, are far from mutually exclusive—is that such characteristics may also have evolved and withstood the test of time for precisely the opposite reason: for their catalytic capacity to actually instigate conflict.

Such a position would better align itself with a more orthodox reading of the evolution of psychopathy. Traditionally, the Darwinian account of psychopathy has rested predominantly on the nonconformist aspect of the disorder (criterion 1, if you recall from chapter 2, for Antisocial Personality Disorder, reads, "Failure to conform to social norms"): on the psychopath's devil-may-care attitude toward social conventions. Conventions, on the one hand, such as honesty, accountability, responsibility, and monogamy*; yet also, on the other hand, conventions like social compliance, which, deep in our ancestral past, would undoubtedly have contributed to perilously poor decision making—and thereby, in those turbulent, treacherous times, to a grisly carnivorous death.

It's the principle of David and Goliath: the little guy with a slingshot lodging his cool, dissenting pebble in the cogs of the all-conquering machine, immune to the pressures of a toxic insider empathy. The lone voice in the wilderness.

Jack the Stripper

Researchers and clinicians often argue that psychopaths don't "do" empathy—that because of their lethargic amygdalae they just don't feel things in the same way as the rest of us. Studies have revealed that when psychopaths are shown distressing images of, say, famine victims, the lights located in the emotion corridors of their brains quite simply don't come on: that their brains—if viewed under fMRI conditions—merely pull down the emotional window blinds and implement a neural curfew.

Sometimes, as we've seen, such curfews can have their advantages—as in the medical profession, for instance. But sometimes the curtains can shut out the light completely. And the darkness can be truly impenetrable.

In the summer of 2010, I jumped on a plane to Quantico, Virginia, to interview Supervisory Special Agent James Beasley III at the FBI's Behavioral Analysis Unit. Beasley is one of America's foremost authorities on psychopaths and serial murder, and has profiled crim-

* Of course, such shameless disregard for the practice of monogamy leads, in turn, to sexual promiscuity . . . and a wider propagation of genes.

inals right across the board: from child abductors to rapists, from drug barons to spree killers.

During his twenty-seven years as a federal employee, the last seventeen of which he's spent at the National Center for the Analysis of Violent Crime, there's not much Special Agent Beasley hasn't heard, seen, or dealt with. But several years ago, he interviewed a guy who was so far down the temperature scale he almost cracked the thermometer.

"There was a string of armed robberies," Beasley explains. "And whoever was behind them wasn't too worried about pulling the trigger. Usually when you deal with armed robberies, the person committing them will just use the gun as a threat.

"But this guy was different. And it was always at close range. A single shot to the head. I was in no doubt whatsoever that we had a psychopath on our hands. The guy was as cold as ice. Mesmerizingly ruthless. But there was something about him that just didn't quite add up. Something about him that bothered me.

"After one of the killings—which, as it happened, turned out to be his last—we caught up with him shortly after that—he'd taken his victim's jacket. Now, that just didn't make sense. Normally, when someone removes an article of clothing from a murder scene, it means one of two things. There's either some kind of sexual stuff going on or there's some other kind of fantasy world being spun out. It's known in the trade as a trophy killing. But neither of those two scenarios fitted this guy's profile. He was too . . . I don't know . . . functional. All business, if you know what I mean.

"So when we brought him in, we asked him. What was the deal with the guy's jacket? And you know what he said? He said, 'Oh, that? It was just a spur-of-the-moment thing. As I was making my way out the door, I looked at the dude as he lay there slumped over the counter. And I suddenly thought to myself, "Hmmh, that jacket kind of goes with my shirt." So what the hell? I thought. The guy's already dead. He's not going anywhere. So I took it. Wore it out to a bar that night, as it happens. Got laid, in fact. You could say it's my lucky jacket. Unlucky for him. But lucky for me . . .' "

When you hear stories like this, it's hard to believe that psychopaths have even heard of empathy, let alone experienced it. Yet, surprisingly, the picture is far from clear in this regard. Mem Mahmut, for instance, showed us that under some sets of circumstances, psychopaths, in fact, seem to be *more* empathic than the rest of us. Or more helpful, at any rate. Then, if you recall, there was the study by Shirley Fecteau and her colleagues, which showed that psychopaths appear to have more going on in their mirror neuron systems, particularly the neurons in the somatosensory areas of their brains—the ones that allow us to identify with others when they're in physical pain—than non-psychopaths.

Whether it's the case that some psychopaths have more empathy than others, or that some are better able to switch it on and off than others, or that some are simply better at faking it than others, is, at present, unknown. But it's a fascinating question that cuts right to the heart of the psychopath's true identity. And one, no doubt, that will be hotly debated for years.

On precisely this issue, I ask Beasley about serial killers. How, in his experience, do *they* measure up on the empathy scale? I'm pretty confident I already know the answer. But Beasley, it turns out, has a surprise in store for me.

"You know, this idea that serial killers lack empathy is a little bit misleading," he says. "Sure, you get your Henry Lee Lucas kind of killer who says that killing a person is just like squashing a bug.* And

*Henry Lee Lucas was a prolific American serial killer, once described as "the greatest monster who ever lived," whose confessions led police to the bodies of 246 victims, 189 of whom he was subsequently convicted of murdering. Lucas's killing spree spanned three decades, from 1960, when he stabbed his mother to death in an argument before having sex with her corpse, to his arrest in 1983, for the unlawful possession of a firearm. In the late 1970s, Lucas teamed up with an accomplice, Ottis Toole, and together the pair drifted around the southern United States, preying primarily, but not exclusively, on hitchhikers. On one occasion, they apparently drove across two states before realizing that the severed head of their latest victim was still on the backseat of their car. "I had no feelings for the people themselves, or any of the crimes," Lucas once stated. "I'd pick them up hitchhiking, running, and playing, stuff like 'at. We'd get to going and having a good time. First thing you

for this functional, instrumental species of serial murderer, perennial drifters who are just after a quick buck, a lack of empathy may well be beneficial, may well contribute to their elusiveness. Dead men tell no tales, right?

"But for another category of serial killer, those we call sadistic serial killers, for whom murder is an end in itself, the presence of empathy—enhanced empathy, even—serves two important purposes.

"Take Ted Bundy, for instance. Bundy ensnared his victims, all of whom were female college students, by pretending to be disabled in some way or other. Arm in a sling, crutches, that kind of thing. Bundy knew, rationally at least, which buttons to press in order to get their assistance. In order to gain their trust. Now, if he hadn't known that, if he hadn't been able to put himself in their shoes, would he really have been able to dupe them so effectively?

"The answer, I believe, is no—a certain degree of cognitive empathy, a modicum of 'theory of mind,' is an essential requirement for the sadistic serial killer.

"On the other hand, however, there has to be a degree of emotional empathy, too. Otherwise how would you derive any enjoyment from watching your victims suffer? From beating them and torturing them and so on? The answer, quite simply, is: you wouldn't.

"So the bottom line, strange though it may seem, is this. Sadistic serial killers feel their victims' pain in exactly the same way that you or I might feel it. They feel it cognitively and objectively. And they feel it emotionally and subjectively, too. But the difference between them and us is that they commute that pain to their own subjective *pleasure*.

"In fact, it would probably be true to say that the greater the amount of empathy they have, the greater the amount of pleasure that they get. Which, when you think about it, is kind of weird."

It most certainly is. But as I sit listening to Beasley, I begin to make connections. Suddenly things start to make sense.

Greg Morant, one of the world's most ruthless con men and a

---

know, I'd killed her and throwed her out somewhere." In 2001, Lucas died in prison, of heart failure. His story is told in the 1986 film *Henry: Portrait of a Serial Killer.*

certified, bona fide psychopath—he simply oozed empathy. That was what made him so good: so mercilessly adept at pinpointing, and zoning in on, his victims' psychological pressure points.

In the mirror neuron study conducted by Shirley Fecteau, psychopaths showed greater empathy than non-psychopaths. The video she showed depicted a scene of physical pain: a needle going into a hand.

And then, of course, there was Mem Mahmut's helping experiment. The fact that the psychopaths managed to out-empathize the non-psychopaths when it came to the "broken arm" condition might have raised his eyebrows, perhaps.

But certainly not James Beasley's.

"Exactly as I'd have predicted," he comments without hesitation. "Though I guess"—he pauses briefly as he weighs up the options— "it kind of depends on which type of psychopath he was testing."

Beasley tells me about a study conducted by Alfred Heilbrun, a psychologist at Emory University, back in the 1980s. Heilbrun analyzed the personality structures of more than 150 criminals and, on the basis of that analysis, differentiated between two very different types of psychopaths: those who had poor impulse control, low IQ, and little empathy (the Henry Lee Lucas type); and those who had better impulse control, high IQ, a sadistic motivation, and heightened empathy (the Ted Bundy or, if you like, Hannibal Lecter type).

But the data concealed a spine-chilling twist. The group, in fact, that exhibited the most empathy of all, according to Heilbrun's taxonomy, comprised high-IQ psychopaths with a history of extreme violence. And in particular, rape: an act that occasionally incorporates a vicarious, sadistic component. Not only are violent acts that inflict pain and suffering on others often more intentional than impulsive, Heilbrun pointed out, echoing Beasley's earlier observation; it is, in addition, precisely through the presence of empathy, and the perpetrator's awareness of the pain being experienced by his victim, that preliminary arousal, and the subsequent satisfaction of sadistic objectives, are achieved.

Not all psychopaths, it would seem, are color-blind. Some see the stop sign in exactly the same way as the rest of us. It's just that they choose to run the light.

The Mask Behind the Face

The fact that a proportion of psychopaths, at least, would appear to experience empathy, and perhaps experience it to a greater degree than the rest of us, may well go some way to clearing up a mystery: how the psychopaths in Angela Book's "vulnerability" study managed to pick up those cues in deportment, those telltale signs in the gait of traumatized assault victims, better than the rest of us.

But if you think psychopaths are alone in their ability to detect splinters of deep emotion invisible to the naked eye, shards of unprocessed feeling buried way beneath the seam of conscious censorship, then you're wrong. Paul Ekman, at the University of California, Berkeley, reports that two Tibetan monks expert in meditation have outperformed judges, policemen, psychiatrists, customs officials, and even Secret Service agents on a subliminal face-processing task that had, up until the monks entered the lab, managed to stump everyone who'd had a go at it (and there were more than five thousand of them).

The task comprises two parts. First, images of faces displaying one of the six basic emotions (anger, sadness, happiness, fear, disgust, and surprise) are flashed on a computer screen. The faces appear for sufficient duration to enable the brain to process them, but insufficient duration for volunteers to be able to consciously report what they see. In the second part of the task, volunteers are required to pick out the face that previously flashed on the screen from a subsequently presented "identity parade" of six.

Typically, performance is at chance level. Over a series of trials, volunteers average a hit rate of around one in six. But the monks averaged three or four. Their secret, Ekman speculates, may hinge on an enhanced, almost preternatural ability to read microexpressions: those minuscule millisecond stroboscopic frames of emotion we learned about earlier that begin to download on the muscles across our face before our conscious brain has time to hit the delete button and display instead the image we wish to present. If so, they'll share that ability with psychopaths. Sabrina Demetrioff, at the University of British Columbia, has recently found precisely such abilities in individuals scoring high on the Hare Self-Report Scale of psychopathy—especially when it comes to expressions of fear and sadness.

Even more intriguing is what happened when Ekman brought one of the monks he'd tested down to the Berkeley Psychophysiology Laboratory, run by his colleague Robert Levenson, to assess his "presence of mind." Here, after being wired up to equipment sensitive to even the minutest fluctuations in autonomic function—muscle contraction, pulse rate, perspiration, and skin temperature—the monk was told that at some point during the ensuing five-minute period he would be subjected to the sound of a sudden loud explosion (the equivalent, Ekman and Levenson decided under the circumstances, of a gun being fired just centimeters away from the ear: the maximal threshold of human acoustical tolerance).

Forewarned of the blast, the monk was instructed to attempt, to the best of his ability, to suppress the inevitable "startle response," to the extent of rendering it, if at all possible, completely imperceptible.

Of course, Ekman and Levenson had been round far too many lab blocks to expect a miracle. Of the hundreds of subjects who had previously filed through the doors, not one of them had managed to flatline. Not even elite police sharpshooters. To not respond at all was impossible. The monitors always picked up something.

Or so they thought.

But they had never tested a Tibetan master of meditation before. And much to their amazement, they finally met their match. Seemingly against all the laws of human physiology, the monk exhibited not the slightest reaction to the explosion. He didn't jump. He didn't flinch. He didn't do anything. He flatlined. The gun went off and the monk just sat there. Like a statue. In all their years, Ekman and Levenson had never seen anything like it.

"When he tries to repress the startle, it almost disappears," Ekman observed afterward. "We've never found anyone who can do that. Nor have any other researchers. This is a spectacular accomplishment. We don't have any idea of the anatomy that would allow him to suppress the startle reflex."

The monk himself, who at the moment of the blast had been practicing a technique known as open presence meditation, put a different slant on it.

"In that state," he explains, "I was not actively trying to control the startle. But the detonation seemed weaker, as if I were hearing it from a distance . . . In the distracted state, the explosion suddenly brings you back to the present moment and causes you to jump out of surprise. But while in open presence you are resting in the present moment. And the bang simply occurs and causes only a little disturbance, like a bird crossing the sky."

I wonder if they tested his hearing.

Roadkill

The work of Paul Ekman, Robert Levenson, and Richard Davidson, mentioned earlier, supports the general consensus that both the cultivation and maintenance of a relaxed state of mind can considerably aid, not just our responses to, but also our perceptions of, the stressors of modern living. Few of us, of course, will ever attain the rarefied spiritual peaks of a Tibetan Buddhist monk. But nearly all of us, on the other hand, will have benefited from keeping a cool head at one time or another.

Yet psychopaths, it would seem, are the exceptions to the rule. In fact, psychopaths, rather than engaging (as do Buddhist monks) in meditation to assimilate inner calm, appear, as their performance on the moral dilemma task demonstrated earlier, to instead have a natural talent for it. And it's not just their results on cognitive decision-making tasks that provide support for such a conclusion. Further evidence of this innate gift for cool comes from basic low-level studies of emotional reactivity.

In work reminiscent of the Emotional Interrupt study mentioned earlier, Chris Patrick of Florida State University compared the reactions of psychopaths and non-psychopaths as they viewed a series of horrific, nauseating, and pleasurable images respectively. On *all* physiological measures—blood pressure, sweat production, heart rate, and blink rate—he found that the psychopaths exhibited significantly less arousal than did normal members of the population. Or, to use the proper terminology, they had an attenuated emotional startle response.

The greatest worth, wrote the eleventh-century Buddhist teacher Atisha, is self-mastery. The greatest magic, transmuting the passions. And to some extent, it would seem, psychopaths have the jump on the rest of us.

But this "jump" isn't always metaphorical in nature. The notion of the psychopath being "one step ahead" can sometimes be just as true in the literal sense of that phrase—as in when getting from point A to point B—as it is when it comes to our responses to emotional stimuli.

And such perpetual peregrination can be just as ascetically demanding. The transient, peripatetic lifestyle, a core feature of the psychopathic personality, has, just like the transmutation of the passions, ancient foundations in the lore of spiritual enlightenment. In Atisha's time, for example, the embodiment of the spiritual archetype was the *shramana*, or wandering monk—and the shramanic ideal of renunciation and abandonment, of loneliness, transience and contemplation, emulated the path to enlightenment followed by the Buddha himself.

These days, of course, the *shramana* is spiritually extinct: a primordial ghost that haunts the star-swept crossroads of spectral, nirvanic wastelands. But in the neon-lit shadows of bars, motels, and casinos, the psychopath is still going strong, assuming, just like his monastic forebears, an itinerant, nomadic existence.

Take serial murder, for example. The latest FBI crime figures estimate that there are around thirty-five to fifty serial killers operating at any one time within the United States. That's a hell of a lot of serial killers by anyone's standards. But dig a little deeper into why this might be the case, and you soon begin to wonder if there shouldn't, in fact, be more.

The U.S. interstate highway system is a schizophrenic beast. During the daytime its rest areas are busily frequented, and have a convivial family vibe. But during the hours of darkness, the mood can quickly change. Many become the haunts of drug dealers and prostitutes on the lookout for easy pickings: long-haul truck drivers and other itinerant workers.

These women aren't exactly missed by their families back home. Many lie dead for weeks, sometimes years, at rest stops and on vacant lots the length and breadth of America, often hundreds of miles away from where they were originally picked up. Police recently discovered the five-to-ten-year-old remains of one of the victims of the Long Island serial killer, for instance, who at the time of writing has been linked to a total of ten murders over a fifteen-year period. The true number of lives claimed by Henry Lee Lucas will never be known. The vastness of the country, the paucity of witnesses, the fact that each state constitutes an independent legal jurisdiction, and the way both victims and offenders are frequently "just passing through" all add up to a logistical, statistical nightmare for the investigating authorities concerned.

I ask one FBI special agent whether he thinks that psychopaths are specially suited to certain types of professions.

He shakes his head.

"Well, they definitely make good truckers." He chuckles. "In fact, I'd even go so far as to say that a truck probably constitutes the most important piece of equipment in the serial killer's toolkit here in the U.S. It's a modus operandi and getaway vehicle rolled into one."

The agent in question is part of a team of law enforcement officers currently working on the FBI's Highway Serial Killings Initiative, a scheme designed to both facilitate the flow of data within America's complex mosaic of autonomous legal jurisdictions and increase public awareness of the murders.

The initiative started almost by accident. In 2004, an analyst from the Oklahoma State Bureau of Investigation detected a pattern. The bodies of murdered women had begun turning up at regular intervals along the Interstate 40 corridor running though Oklahoma, Texas, Arkansas, and Mississippi. Analysts working on the Violent Criminal Apprehension Program (ViCAP), a national matrix containing information on homicides, sexual assaults, missing persons, and unidentified human remains, scanned their database to see if similar patterns of highway killings existed elsewhere.

They did—and then some. So far, ongoing investigations have revealed more than five hundred murder victims from along or near highways, as well as a list of some two hundred potential suspects.

"Psychopaths are shadowmancers," the agent tells me, a large-scale map of the United States dotted with timelines, hot spots, and murderous crimson trajectories plastered across the wall behind his desk. "They survive by moving around. They don't have the same need for close relationships that normal people do. So they live in an orbit of perpetual drift, in which the chances of running into their victims again is minimized.

"But they can also turn on the charm. Which, in the short or medium term, at least, allows them to stay in one place for a sufficient length of time to allay suspicion—and cultivate victims. This extraordinary charisma—and it borders, in some cases, on the supernatural: even though you know they're cold as ice, and would kill you as soon as look at you, you sometimes just can't help liking them—acts as a kind of psychological smoke screen that masks their true intentions.

"This, by the way, is also why you generally tend to find a higher proportion of psychopaths in urban, as opposed to rural, areas. In a city, anonymity's easy to come by. But you just try melting into the crowd in a farming or a mining community. You're going to have a hard job.

"Unfortunately, the words 'psychopath' and 'drifter' go together hand in glove. And that's a huge headache for law enforcement agencies. That, right there, is what makes our job so goddamned difficult at times."

The Lesson of the Moth

Peter Jonason, the purveyor of "James Bond" psychology, has a theory about psychopathy. Exploiting others is a highly risky business, he points out, which often results in failure. Not only are people on the lookout for cutthroats and shysters, they are also inclined to react badly to them, legally or otherwise. If you're going to cheat, Jonason elucidates, being extroverted, charming, and high in self-esteem make it easier to cope with rejection. And easier to hit the road.

Bond, of course, was always on the road. As a spy it comes with the territory, same as it does for the serial killer out on the interstate, same as it did for the wandering monks of old. But although these three have rather different reasons for their travels and occupy rather different stations along the psychopathic spectrum, they are also guided by a common metaphysical blueprint—the ceaseless quest for novel, heightened experience, be it a fight to the death with a deranged criminal mastermind; the unfathomable, toxic power in taking another's life; or the transcendental purity of eternal peregrination.

Such openness to experience is a quality common to both psychopaths and saints, and, if you recall, constitutes an integral component of mindfulness meditation. Yet it is one of quite a number that these two apparent opposites happen to share (see figure 7.2 below). Not all psychopathic traits are spiritual traits, and vice versa. But there are some, as we've seen, that undoubtedly overlap, of which openness to experience is perhaps the most fundamental. Hunter S. Thompson would certainly agree. And it is, after all, the only one we're born with.

After wrapping things up with the FBI in Quantico, I meander down to Florida for a holiday. Killing some time in downtown Miami before catching my flight home, I chance, on a cloudless Sunday morn-

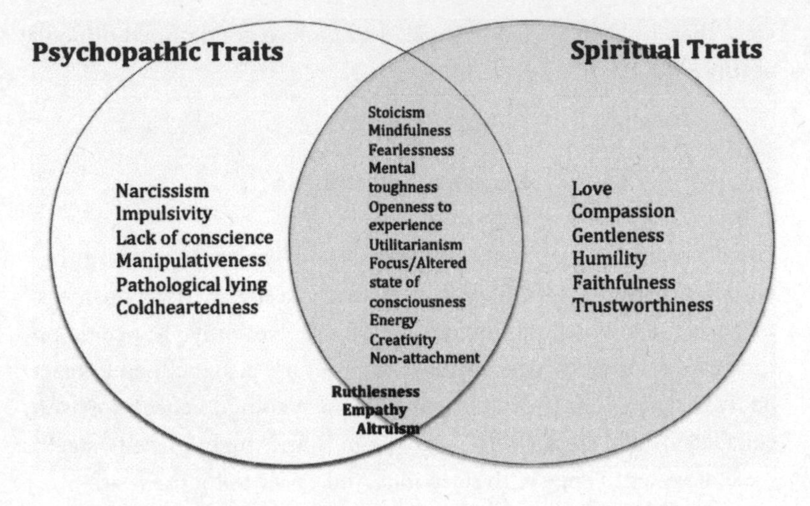

Figure 7.2. The relationship between psychopathic and spiritual states

ing in the *calles* of Little Havana, upon a flea market. On a table of bric-a-brac, by the side of a stack of jigsaw puzzles, is a copy of *Archy and Mehitabel*, its midnight-blue dust jacket sandblasted a tropical turquoise by the sun and the salt of the ocean. The book, originally penned in 1927 by the celebrated New York columnist Don Marquis, chronicles the verse of its unlikely title character, Archy—a cockroach auteur with a peculiar penchant for poetry—and his oddball adventures with best pal Mehitabel, a reincarnated alley cat, who claims, in a previous life, to have once been Cleopatra.

I thumb through the pages. And cough up a couple of dollars. It'll do for the plane back, I think. Later that night, forty thousand feet over the slumbering North Atlantic, I come across the following poem.

It's a poem about moths. But it's also a poem about psychopaths.

I get it copied. And stick it in a frame. And now it glowers redoubtably above my desk. An entomological keepsake of the horizons of existence. And the brutal, star-crossed wisdom of those who seek them out.

i was talking to a moth
the other evening
he was trying to break into
an electric light bulb
and fry himself on the wires

why do you fellows
pull this stunt i asked him
because it is the conventional
thing for moths or why
if that had been an uncovered
candle instead of an electric
light bulb you would
now be a small unsightly cinder
have you no sense

plenty of it he answered
but at times we get tired
of using it
we get bored with the routine
and crave beauty
and excitement
fire is beautiful
and we know that if we get
too close it will kill us
but what does that matter
it is better to be happy
for a moment
and be burned up with beauty
than to live a long time
and be bored all the while
so we wad all our life up
into one little roll
and then we shoot the roll

that is what life is for
it is better to be a part of beauty
for one instant and then cease to
exist than to exist forever
and never be a part of beauty
our attitude toward life
is come easy go easy
we are like human beings
used to be before they became
too civilized to enjoy themselves

and before i could argue him
out of his philosophy
he went and immolated himself
on a patent cigar lighter
i do not agree with him
myself i would rather have
half the happiness and twice
the longevity

but at the same time i wish
there was something i wanted
as badly as he wanted to fry himself

NOTES

Preface

xii *"They say that humans developed fear as a survival mechanism . . ."* See Arne Öhman and Susan Mineka, "The Malicious Serpent: Snakes as a Prototypical Stimulus for an Evolved Module of Fear," *Current Directions in Psychological Science* 12, no. 1 (2003): 5–9, doi:10.1111/1467-8721.01211. For an easy-to-read introduction to the evolutionary basis of emotion, see Joseph E. LeDoux, *The Emotional Brain: The Mysterious Underpinnings of Emotional Life* (New York: Simon & Schuster, 1996).

xii *Monkeys with lesions of the amygdala . . .* See Heinrich Klüver and Paul C. Bucy, "'Psychic Blindness' and Other Symptoms Following Bilateral Temporal Lobectomy in Rhesus Monkeys," *American Journal of Physiology* 119 (1937): 352–53; Klüver and Bucy, "Preliminary Analysis of Functions of the Temporal Lobes in Monkeys," *Archives of Neurology and Psychiatry* 42, no. 6 (1939): 979–1000.

xii *"There was no such thing as stock in the Pleistocene era . . ."* Quote taken from Jane Spencer, "Lessons from the Brain-Damaged Investor," *The Wall Street Journal*, July 21, 2005. http://online.wsj.com/article/0,,SB112190164023291519,00.html.

xiii *Even today, anxious individuals are better than the rest of us . . .* See Elaine Fox, Riccardo Russo, and George A. Georgiou, "Anxiety Modulates the Degree of Attentive Resources Required to Process Emotional Faces," *Cognitive, Affective, and Behavioral Neuroscience* 5, no. 4 (2005): 396–404, doi:10.3758/CABN.5.4.396.

xiii *In his book* The Man Who Mistook His Wife for a Hat . . . Oliver Sacks, *The Man Who Mistook His Wife for a Hat: And Other Clinical Tales* (New York: Summit Books, Simon & Schuster, 1985).

xiii *Kéri has found that people . . .* See Szabolcs Kéri, "Genes for Psychosis and Creativity: A Promoter Polymorphism of the Neuregulin 1 Gene Is Related to Creativity in People with High Intellectual Achievement," *Psychological Science* 20, no. 9 (2009): 1070–73, doi: 10.1111/j.1467-9280.2009.02398.x.

xiv *In an ingenious experiment, Joe Forgas . . .* Joseph P. Forgas, Liz Goldenberg, and Christian Unkelbach, "Can Bad Weather Improve Your Memory? An

Unobtrusive Field Study of Natural Mood Effects on Real-Life Memory," *Journal of Experimental Social Psychology* 45 (2009): 254–57, doi:10.1016/j.jesp.2008.08.014.

xv *A psychopath wouldn't worry . . .* Psychopathy is far more prevalent in men than in women. A number of reasons have been put forward as to why this might be the case. Developmental theorists stress that differences in aggression might arise from dichotomous parental socialization practices in the upbringing of boys and girls, while also pointing to the fact that girls show earlier development of linguistic and socio-emotional skills than boys, which may in turn predispose them to the emergence of more effective behavior inhibition strategies. Evolutionary theorists, on the other hand, emphasize hardwired gender differences in behavioral activation and withdrawal as a possible source of the discrepancy: women, for instance, tend to report more "negative withdrawal" emotions (such as fear) in the presence of aversive stimuli, while men report more "negative activation" emotions such as anger. A third school of thought highlights the possible role of sociological factors in the "presence" of the disorder: a subtle gender bias in diagnosis on the part of clinicians, for example, allied with the traditional social stigma attached to women presenting with externalizing antisocial psychopathology, indeed even *reporting* antisocial feelings and attitudes. Whatever the reason, and it's probable that the demographic conceals a confluence of all three factors, estimates of the incidence of psychopathy tend to vary from between 1 to 3 percent in men to 0.5 to 1 percent in women.

1. Scorpio Rising

5 *Back in the 1990s, Hare submitted a research paper . . .* The paper in question, which Hare *did* eventually get published, was the following: Sherrie Williamson, Timothy J. Harpur, and Robert D. Hare, "Abnormal Processing of Affective Words by Psychopaths," *Psychophysiology* 28, no. 3 (1991): 260–73, doi:10.1111/j.1469-8986.1991.tb02192.x.

8 *To find out, she set up a simple experiment . . .* See Sarah Wheeler, Angela Book, and Kimberley Costello, "Psychopathic Traits and the Perception of Victim Vulnerability," *Criminal Justice and Behavior* 36, no. 6 (2009): 635–48, doi:10.1177/0093854809333958. It should also be noted that, while psychopaths may possess a vulnerability radar, there is evidence to suggest that elements of their own body language "leak out" and set them apart from normal members of the population. One study, for instance, using video footage, has shown that psychopaths may be reliably differentiated from non-psychopaths on the basis of as little as five- and ten-second exposure sequences. See Katherine A. Fowler, Scott O. Lilienfeld, and Christopher J. Patrick, "Detecting Psychopathy from Thin Slices of Behavior," *Psychological Assessment* 21, no. 1 (2009): 68–78, doi:10.1037/a0014938.

8 *First, she handed out the Self-Report Psychopathy Scale* . . . See Delroy L. Paulhus, Craig S. Neumann, and Robert D. Hare, *Self-Report Psychopathy Scale: Version III* (Toronto: Multi-Health Systems, in press).

8 *Moreover, when Book repeated the procedure* . . . Kimberley Costello and Angela Book, "Psychopathy and Victim Selection," poster presented at the Society for the Scientific Study of Psychopathy conference, Montreal, Canada, May 2011.

12 *In 2009, shortly after Angela Book published the results of her study, I decided to perform my own take on it.* This is an ongoing study, and further data are currently being collected in order to substantiate these initial findings.

13 *In 2003, Reid Meloy* . . . See J. Reid Meloy and M. J. Meloy, "Autonomic Arousal in the Presence of Psychopathy: A Survey of Mental Health and Criminal Justice Professionals," *Journal of Threat Assessment* 2, no. 2 (2002): 21–33, doi: 10.130015177v02n02-02.

14 *Kent Bailey, emeritus professor in clinical psychology* . . . See Kent G. Bailey, "The Sociopath: Cheater or Warrior Hawk?" *Behavioral and Brain Sciences* 18, no. 3 (1995): 542–43, doi:10.1017/S0140525X00039613.

14 *Robin Dunbar, professor of evolutionary anthropology at Oxford University* . . . See Robin I. M. Dunbar, Amanda Clark, and Nicola L. Hurst, "Conflict and Cooperation among the Vikings: Contingent Behavioral Decisions," *Ethology and Sociobiology* 16, no. 3 (1995): 233–46, doi:10.1016/0162-3095(95)00022-D.

16 *Joshua Greene, a psychologist, neuroscientist, and philosopher* . . . For more on the work of Joshua Greene, and the fascinating interface between neuroscience and moral decision making, see Joshua D. Greene, R. Brian Sommerville, Leigh E. Nystrom, John M. Darley, and Jonathan D. Cohen, "An fMRI Investigation of Emotional Engagement in Moral Judgment," *Science* 293, no. 5537 (2001): 2105–08, doi:10.1126/science.1062872; Andrea L. Glenn, Adrian Raine, and R. A. Schug, "The Neural Correlates of Moral Decision-Making in Psychopathy," *Molecular Psychiatry* 14 (January 2009): 5–6, doi:10.1038/mp.2008.104.

16 *Consider, for example, the following conundrum (case 1)* . . . The Trolley Problem was first proposed in this form by Philippa Foot in "The Problem of Abortion and the Doctrine of the Double Effect," in *Virtues and Vices: And Other Essays in Moral Philosophy* (Berkeley: University of California Press, 1978).

16 *Now consider the following variation (case 2)* . . . See Judith Jarvis Thomson, "Killing, Letting Die, and the Trolley Problem," *The Monist* 59, no. 2 (1976): 204–17.

19 *Daniel Bartels at Columbia University and David Pizarro at Cornell* . . . See Daniel M. Bartels and David A. Pizarro, "The Mismeasure of Morals: Antisocial Personality Traits Predict Utilitarian Responses to Moral Dilemmas," *Cognition* 121, no. 1 (2011): 154–61.

21 *In 2005, Belinda Board and Katarina Fritzon* . . . See Belinda J. Board and Katarina Fritzon, "Disordered Personalities at Work," *Psychology, Crime, and Law* 11, no. 1 (2005): 17–32, doi:10.1080/10683160310001634304.

21 *Mehmet Mahmut and his colleagues at Macquarie University* . . . See Mehmet K. Mahmut, Judi Homewood, and Richard J. Stevenson, "The Characteristics of Non-Criminals with High Psychopathy Traits: Are They Similar to Criminal Psychopaths?" *Journal of Research in Personality* 42, no. 3 (2008): 679–92.

21 *In a similar (if less high-tech) vein* . . . Unpublished pilot survey.

22 *Jon Moulton, one of London's most successful venture capitalists* . . . See Emma Jacobs, "20 Questions: Jon Moulton," *Financial Times*, February 4, 2010, www.ft.com/cms/s/0/32c642f2-11c1-11df-9d45-00144feab49a.html#axzz1srPuKoUq.

22 *But there's a story I once heard* . . . For this story, I would like to thank Nigel Henbest and Heather Couper.

24 *Back in the 1980s, Harvard researcher Stanley Rachman* . . . For more on Rachman's work see Stanley J. Rachman, "Fear and Courage: A Psychological Perspective," *Social Research* 71, no. 1 (2004): 149–76. Rachman makes it quite clear in this paper that bomb disposal experts are *not* psychopathic—a view echoed here. Rather, the point being made is that confidence and coolness under pressure are two traits that psychopaths and bomb-disposal experts have in common.

25 *Relationship experts Neil Jacobson and John Gottman* . . . *have observed* . . . See Neil Jacobson and John Gottman, *When Men Batter Women: New Insights into Ending Abusive Relationships* (New York: Simon & Schuster, 1998).

26 *In 2009, Lilianne Mujica-Parodi* . . . See Lilianne R. Mujica-Parodi, Helmut H. Strey, Frederick Blaise, Robert Savoy, David Cox, Yevgeny Botanov, Denis Tolkunov, Denis Rubin, and Jochen Weber, "Chemosensory Cues to Conspecific Emotional Stress Activate Amygdala in Humans," *PLoS ONE* 4, no. 7 (2009): e6415, doi:10.1371/journal.pone.0006415.

27 *To find out, I ran a variation on Mujica-Parodi's study* . . . Paper submitted for publication. It should be noted, with regard to my own study, that the psychopaths were no better than the non-psychopaths at detecting which was the fear sweat and which was the non-fear sweat on the basis of odor. The distinctive odor of any sweat comes from bacterial contamination, and the collection and storage protocols were designed, as in the Mujica-Parodi study, to prevent bacterial growth. The difference between the psychopaths and the non-psychopaths was in the effect that exposure to the fear sweat had on performance.

2. Will the Real Psychopath Please Stand Up?

35 *Then, in 1952, the British psychologist Hans Eysenck* . . . For an in-depth treatment of Eysenck's contribution to personality theory, see Hans J. Eysenck and Michael W. Eysenck, *Personality and Individual Differences: A Natural Science Approach* (New York: Plenum Press, 1985). For the original paper incorporating Hippocrates' four temperaments, see Hans J. Eysenck, "A Short Question-

naire for the Measurement of Two Dimensions of Personality," *Journal of Applied Psychology* 42, no. 1 (1958): 14–17.

36 *Eysenck's two-stroke model of personality was positively anorexic* . . . See Gordon W. Allport and Henry S. Odbert, "Trait-Names: A Psycho-Lexical Study," *Psychological Monographs* 47, no. 1 (1936): i–171, doi:10.1037/h0093360.

37 *But it wasn't until University of Illinois psychologist Raymond Cattell* . . . See Raymond B. Cattell, *The Description and Measurement of Personality* (New York: Harcourt, Brace, and World, 1946) and Cattell, *Personality and Motivation: Structure and Measurement* (Yonkers-on-Hudson, NY: World Book Co., 1957).

38 *In 1961, two U.S. Air Force researchers, Ernest Tupes and Raymond Christal* . . . See Ernest C. Tupes and Raymond E. Christal, "Recurrent Personality Factors Based on Trait Ratings," Technical Report ASD-TR-61-97, Personnel Laboratory, Aeronautical Systems Division, Air Force Systems Command, United States Air Force, Lackland Air Force Base, Texas, May 1961. Republished in *Journal of Personality* 60, no. 2 (1992): 225–51, doi:10.1111/j.1467-6494.1992.tb00973.x.

38 *More recently, over the last twenty years or so* . . . See Paul T. Costa and Robert R. McCrae, "Primary Traits of Eysenck's P-E-N System: Three- and Five-Factor Solutions," *Journal of Personality and Social Psychology* 69, no. 2 (1995): 308–17.

38 *Psychologists don't really do consensus if they can help it* . . . So indivisible are the Big Five atoms of personality that they've even been observed across species. A 1997 study, conducted by James King and Aurelio Figueredo at the University of Arizona, reveals that the chimpanzee personality also conforms to the five-factor model found in humans—with, it turns out, one extra thrown in for good measure: dominance, an evolutionary artifact of hierarchical chimp society. Sam Gosling, now at the University of Texas at Austin, has performed similar work with hyenas. Gosling recruited four volunteers to provide, with the help of specially designed scales, standardized personality ratings of a group of *Crocuta crocuta* (a species of spotted hyena). The hyenas were housed at the Field Station for Behavioral Research at the University of California, in Berkeley. Lo and behold, when Gosling went over the data, five dimensions leapt from the spreadsheet in front of him: Assertiveness; Excitability; Human-Directed Agreeableness; Sociability; and Curiosity. Which, if we forget about Conscientiousness for a moment, make a pretty good fit with the remaining four contenders (Neuroticism; Agreeableness; Extraversion; and Openness to Experience). And Gosling didn't stop there. Encouraged by these results, he took his emotional calculus to the seabed—and discovered clear differences in sociability . . . in octopuses. Some octopuses, it appears, prefer to eat in the safety of their own dens, while others like dining alfresco. See James E. King and Aurelio J. Figueredo, "The Five-Factor Model plus Dominance in Chimpanzee Personality," *Journal of Research in Personality* 31 (1997): 257–71; Samuel D. Gosling, "Personality Dimensions in Spotted Hyenas (*Crocuta*

crocuta)," *Journal of Comparative Psychology* 112, no. 2 (1998): 107–118. Also, for a more general look at personality traits in the animal kingdom at large, see S. D. Gosling and Oliver P. John, "Personality Dimensions in Nonhuman Animals: A Cross-Species Review," *Current Directions in Psychological Science* 8, no. 3 (1999): 69–75, doi:10.1111/1467-8721.00017.

38 *Or, as they're commonly referred to, the "Big Five"*... For more on the structure of personality, and in particular the Big Five, see R. R. McCrae and P. T. Costa, *Personality in Adulthood* (New York: Guilford Press, 1990); McCrae and Costa, "A Five-Factor Theory of Personality," in Lawrence A. Pervin and O. P. John (eds.), *Handbook of Personality: Theory and Research*, 2nd ed. (New York: Guilford Press, 1999), 139–53.

40 *In doing so, they've found a striking connection*... For more on the relationship between personality and career choice, see Adrian Furnham, Liam Forde, and Kirsti Ferrari, "Personality and Work Motivation," *Personality and Individual Differences* 26, no. 6 (1999): 1035–43; A. Furnham, Chris J. Jackson, L. Forde, and Tim Cotter, "Correlates of the Eysenck Personality Profiler," *Personality and Individual Differences* 30, no. 4 (2001): 587–94.

41 *Lynam and his colleagues at the University of Kentucky* . . . For the Lynam study, see Joshua D. Miller, Donald R. Lynam, Thomas A. Widiger, and Carl Leukefeld, "Personality Disorders as Extreme Variants of Common Personality Dimensions: Can the Five-Factor Model Adequately Represent Psychopathy?" *Journal of Personality* 69, no. 2 (2001): 253–76. For more on the relationship between psychopathy and the five-factor model of personality, see T. A. Widiger and D. R. Lynam, "Psychopathy and the Five Factor Model of Personality," in Theodore Millon, Erik Simonsen, Morten Birket-Smith, and Roger D. Davis (eds.), *Psychopathy: Antisocial, Criminal, and Violent Behavior* (New York: Guilford Press, 1998): 171–87; and J. D. Miller and D. R. Lynam, "Psychopathy and the Five-Factor Model of Personality: A Replication and Extension," *Journal of Personality Assessment* 81, no. 2 (2003): 168–78. For an analysis of the relationship between the five-factor model and other personality disorders, including psychopathy, see P. T. Costa and R. R. McCrae, "Personality Disorders and the Five-Factor Model of Personality," *Journal of Personality Disorders* 4, no. 4 (1990): 362–71, doi:10.1521/pedi.1990.4.4.362.

42 *The picture of a U.S. president?* See Scott O. Lilienfeld, Irwin D. Waldman, Kristin Landfield, Ashley L. Watts, Steven J. Rubenzer, and Thomas R. Faschingbauer, "Fearless Dominance and the U.S. Presidency: Implications of Psychopathic Personality Traits for Successful and Unsuccessful Political Leadership," *Journal of Personality and Social Psychology* (Epub, abstract posted July 23, 2012, doi: 10.1037/a0029392).

42 *Back in 2000, Rubenzer and Faschingbauer had sent out* . . . See S. J. Rubenzer, T. R. Faschingbauer, and Deniz S. Ones, "Assessing the U.S. Presidents Using the Revised NEO Personality Inventory," in "Innovations in Assessment with the Revised NEO Personality Inventory," ed. R. R. McCrae and

P. T. Costa, special issue, *Assessment* 7, no. 4 (2000): 403–19, doi:10.1177/1073191 10000700408. For more on the development and structure of the NEO Personality Inventory, see: P. T. Costa and R. R. McCrae, *Revised NEO Personality Inventory (NEO-PI-R) and NEO Five-Factor Inventory (NEO-FFI) Professional Manual* (Odessa, FL: Psychological Assessment Resources, 1992); Costa and McCrae, "Domains and Facets: Hierarchical Personality Assessment Using the Revised NEO Personality Inventory," *Journal of Personality Assessment* 64, no. 1 (1995): 21–50.

44 *DSM classifies personality disorders* . . . The complete inventory of component disorders comprising each cluster can be found at www.wisdomofpsychopaths.com.

44 *Support for this latter, anti-separationist view* . . . See Lisa M. Saulsman and Andrew C. Page, "The Five-Factor Model and Personality Disorder Empirical Literature: A Meta-Analytic Review," *Clinical Psychology Review* 23, no. 8 (2004): 1055–85.

45 *But, crucially, it was an overriding "Big Two" that did most of the heavy lifting* . . . Here are Saulsman and Page's findings in graphical form:

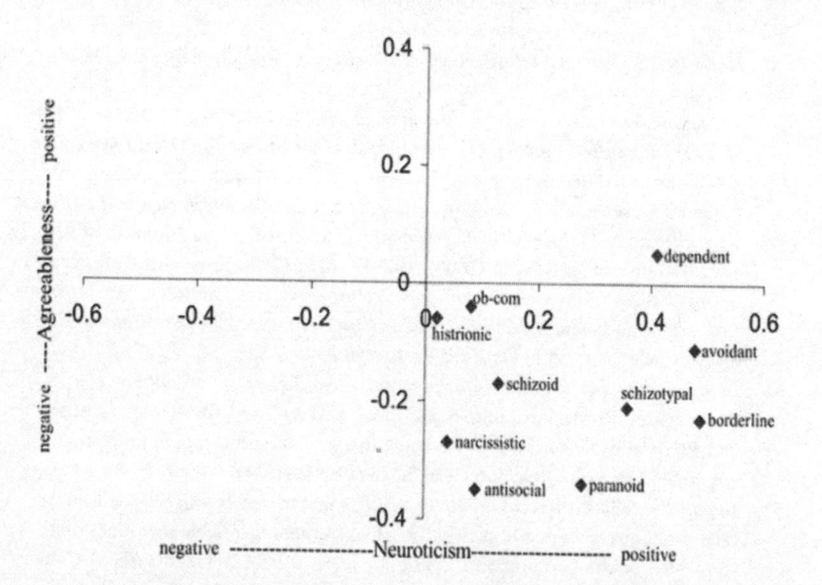

45 *The philosopher Theophrastus* . . . Theophrastus, *Characters*, trans. James Diggle, Cambridge Classical Texts and Commentaries (Cambridge: Cambridge University Press, 2005).

46 *In 1801, a French physician by the name of Philippe Pinel* . . . See Philippe Pinel, *Medico-Philosophical Treatise on Mental Alienation, Second Edition, Entirely*

Reworked and Extensively Expanded (1809), trans. of *Traité médico-philosophique sur l'aliénation mentale, 1809* by Gordon Hickish, David Healy, and Louis C. Charland (Oxford: Wiley-Blackwell, 2008).

46 *The physician Benjamin Rush, practicing in America in the early 1800s . . .* See Benjamin Rush, *Medical Inquiries and Observations upon the Diseases of the Mind* (New York: New York Academy of Medicine, 1812; New York: Hafner, 1962).

47 *In his book* The Mask of Sanity, *published in 1941 . . .* See Hervey Cleckley, *The Mask of Sanity: An Attempt to Clarify Some Issues About the So-Called Psychopathic Personality* (St. Louis, MO: C. V. Mosby, 1941, 1976). The entire text may be downloaded for free at www.cassiopaea.org/cass/sanity_1.pdf.

48 *This moral conundrum was first put forward by Judith Jarvis Thomson . . .* See Judith Jarvis Thomson, "The Trolley Problem," *Yale Law Journal* 94, no. 6 (1985): 1395–1415.

49 *In 1980, Robert Hare (whom we met in chapter 1) unveiled the Psychopathy Checklist . . .* See Robert D. Hare, "A Research Scale for the Assessment of Psychopathy in Criminal Populations," *Personality and Individual Differences* 1, no. 2 (1980): 111–19, doi:10.1016/0191-8869(80)90028-8.

49 *The checklist—which, in 1991, underwent a facelift . . .* See R. D. Hare, *The Hare Psychopathy Checklist Revised: Technical Manual* (Toronto: Multi-Health Systems, 1991).

50 *but recent activity by a number of clinical psychologists . . .* See R. D. Hare, *The Hare Psychopathy Checklist Revised*, 2nd ed. (Toronto: Multi-Health Systems, 2003). For a detailed overview of the dynamical structure of the psychopathic personality, see Craig S. Neumann, R. D. Hare, and Joseph P. Newman, "The Super-Ordinate Nature of the Psychopathy Checklist Revised," *Journal of Personality Disorders* 21, no. 2 (2007): 102–17. Also R. D. Hare and C. S. Neumann, "The PCL-R Assessment of Psychopathy: Development, Structural Properties, and New Directions," in Christopher J. Patrick (ed.), *Handbook of Psychopathy* (New York: Guilford Press, 2006), 58–88.

52 *In prison populations ASPD is the psychiatric equivalent of the common cold . . .* See Megan J. Rutherford, John S. Cacciola, and Arthur I. Alterman, "Antisocial Personality Disorder and Psychopathy in Cocaine-Dependent Women," *American Journal of Psychiatry* 156, no. 6 (1999): 849–56.

52 *In addition, this 20 percent minority punches well above its weight . . .* For more facts and figures on psychopathy, plus an extremely accessible introduction to the world of the psychopath in general, see Robert D. Hare, *Without Conscience: The Disturbing World of the Psychopaths Among Us* (New York: Guilford Press, 1993).

52 *Studies comparing the recidivism rates among psychopathic and non-psychopathic prisoners . . .* See James F. Hemphill, R. D. Hare, and Stephen Wong, "Psychopathy and Recidivism: A Review," *Legal and Criminological Psychology* 3, no. 1 (1998): 139–70, doi:10.1111/j.2044-8333.1998.tb00355.x.

53 *Jimmy is thirty-four years old and has been sentenced to life imprisonment . . .* The idea for two prototypical, fictionalized stories highlighting the differences between psychopathy and Antisocial Personality Disorder, respectively, is borrowed from a similar depiction featured in James Blair, Derek Mitchell, and Karina Blair, *The Psychopath: Emotion and the Brain* (Malden, MA: Blackwell, 2005), 4–6.

56 *For a start, studies reveal that concordance rates . . .* See Hare, *The Hare Psychopathy Checklist Revised* (Toronto: Multi-Health Systems, 1991).

57 *A recent study led by Stephanie Mullins-Sweatt . . .* See Stephanie M. Mullins-Sweatt, Natalie G. Glover, Karen J. Derefinko, Joshua M. Miller, and Thomas A. Widiger, "The Search for the Successful Psychopath," *Journal of Research in Personality* 44, no. 4 (2010): 554–58.

60 *The Psychopathic Personality Inventory (PPI for short) . . .* See Scott O. Lilienfeld, and Brian P. Andrews, "Development and Preliminary Validation of a Self-Report Measure of Psychopathic Personality Traits in Noncriminal Populations," *Journal of Personality Assessment* 66, no. 3 (1996): 488–524.

61 *The autistic spectrum, for instance . . .* Autistic spectrum disorders include autism, Asperger's syndrome, childhood disintegrative disorder, Rett syndrome, and pervasive developmental disorder not otherwise specified. For more information on autism in general, go to: www.autism.org.uk/. For more information on the idea of an autistic spectrum, go to www.autism.org.uk/about-autism/autism-and-asperger-syndrome-an-introduction/what-is-autism.aspx.

61 *Less familiar, perhaps, but equally pertinent . . .* For more information on schizophrenia—symptoms, diagnosis, treatment, and support—go to www.schizophrenia.com/. For more information on the schizophrenic spectrum, and possible underlying neural correlates, go to www.schizophrenia.com/sznews/archives/002561.html.

64 *In an ingenious experiment, Newman demonstrated . . .* See Kristina D. Hiatt, William A. Schmitt, and Joseph P. Newman, "Stroop Tasks Reveal Abnormal Selective Attention Among Psychopathic Offenders," *Neuropsychology* 18, no. 1 (2004): 50–59.

66 *In a separate study, Newman and his colleagues . . .* See Joseph P. Newman, John J. Curtin, Jeremy D. Bertsch, and Arielle R. Baskin-Sommers, "Attention Moderates the Fearlessness of Psychopathic Offenders," *Biological Psychiatry* 67, no. 1 (2010): 66–70.

3. Carpe Noctem

74 *"Two Police Community Support Officers did not intervene . . ."* Matthew Moore, "Officers 'Not Trained' to Rescue Drowning Boy," *The Telegraph*, September 21, 2007, www.telegraph.co.uk/news/uknews/1563717/Officers-not-trained-to-rescue-drowning-boy.html.

75 *In an experiment, for instance, that hitched the latest in social networking . . .*

See Vladas Griskevicius, Noah J. Goldstein, Chad R. Mortensen, Robert B. Cialdini, and Douglas T. Kenrick, "Going Along Versus Going Alone: When Fundamental Motives Facilitate Strategic (Non)Conformity," *Journal of Personality and Social Psychology* 91, no. 2 (2006): 281–94, doi:10.1037/0022-3514.91.2.281.

75 *The psychologist Irving Janis* . . . See Irving L. Janis and Leon Mann, *Decision Making: A Psychological Analysis of Conflict, Choice and Commitment* (New York: Free Press, 1977).

77 *Andrew Colman, professor of psychology at the University of Leicester* . . . See Andrew M. Colman, Andrew and J. Clare Wilson, "Antisocial Personality Disorder: An Evolutionary Game Theory Analysis," *Legal and Criminological Psychology* 2, no. 1 (1997): 23–34, doi:10.1111/j.2044-8333.1997.tb00330.x.

79 *In 2010, Hideki Ohira, a psychologist at Nagoya University* . . . See Takahiro Osumi and Hideki Ohira, "The Positive Side of Psychopathy: Emotional Detachment in Psychopathy and Rational Decision-Making in the Ultimatum Game," *Personality and Individual Differences* 49, no. 5 (2010): 451–56.

83 *"To subdue the enemy without fighting . . ."* This quote originates in Sun Tzu's much-celebrated work of military strategy, *The Art of War*. See *The Art of War by Sun Tzu—Special Edition*, trans. and ed. Lionel Giles (1910; repr. El Paso, TX: El Paso Norte Press, 2005).

83 *One observes a still similar dynamic in monkeys today* . . . For the latest on altruistic behavior in chimpanzees, see Victoria Horner, J. Devyn Carter, Malini Suchak, and Frans B. M. de Waal, "Spontaneous Prosocial Choice by Chimpanzees," *PNAS* 108, no. 33 (2011): 13847–51, doi:10.1073/pnas.1111088108. Altruistic combat is also observed in birds. Male ravens, for example, compete with each other for mates not by means of aggression, but rather by performing "acts of bravery." That is, instead of going beak to beak in ornithological combat, they challenge each other to deadly games of one-upmanship: the "game," in this case, comprising the hazardous undertaking of verifying whether or not potential carrion is dead (the perilous alternatives being sleeping, injured, or pretending). "By demonstrating that they have the courage, experience, and quickness of reaction to deal with life's dangers," says Frans de Waal, professor of primate behavior at Emory University, "the occasional boldness of corvids serves to enhance status and impress potential mates." (Quote taken from Frans B. M. de Waal, *Good Natured: The Origins of Right and Wrong in Humans and Other Animals* [Cambridge, MA: Harvard University Press, 1996], 134.)

84 *"instead of dominants standing out because of what they take . . ."* Quote taken from de Waal, *Good Natured*, 144.

84 *"on the basis of how best to restore peace"* Ibid., 129.

84 *"the group looks for the most effective arbitrator . . ."* Ibid., 144.

84 *In 1979, at a remote site near the village of Saint-Césaire in southwest France* . . . Christoph P. E. Zollikofer, Marcia S. Ponce de León, Bernard Vandermeersch,

and François Lévêque, "Evidence for Interpersonal Violence in the St. Césaire Neanderthal," *PNAS* 99, no. 9 (2002): 6444–48, doi:10.1073/pnas.082111899.

88 *the whole purpose of the Prisoner's Dilemma . . .* The Prisoner's Dilemma was originally conceived at the RAND Corporation in 1950 by the mathematicians Merrill Flood and Melvin Dresher. Later that same year, the game was first formulated with prison-sentence payoffs by Albert Tucker and given its official title.

88 *Instead, the screen of life is densely populated with millions upon millions of individual pixels . . .* In a world of "repeated interaction" (such as everyday life), a psychopathic strategy does indeed have its shortcomings. Such an observation, however, does not take into consideration the following two points:

A. By moving from place to place, the psychopath, unfettered by the need for close relationships, creates his or her own "virtual world" in which the chance of repeated encounters is minimized.

B. Psychopaths' supreme ability to charm and adopt psychological camouflage safeguards, to some extent, against their identity as "defector" coming out. It acts, in the short or medium term, at least, as a smoke screen, permitting their misdemeanors to go unnoticed. Indeed, avoidance of detection also goes some way to explaining the greater incidence of psychopathy in urban settings—where anonymity, if one so wishes, is assured—as opposed to rural areas, where "melting into the crowd" is less of an option.

Bottom line? Psychopaths have exactly the right "personality kit" to bend, or break, the rules. If you're going to cheat in the game of life, being ruthless and fearless ensures you're never too far out of your comfort zone, while being extroverted and charming can help you get away with it for longer. And, in the event that you *are* found out, high self-esteem makes it easier to cope with rejection.

92 *In the late 1970s, the political scientist Robert Axelrod . . .* For more on the virtual reality tournament set up by Robert Axelrod, and on the precepts of game theory in general, see Robert Axelrod, *The Evolution of Cooperation* (New York: Basic Books, 1984).

94 *Such an abomination had already occurred . . .* See Robert L. Trivers, "The Evolution of Reciprocal Altruism," *Quarterly Review of Biology* 46, no. 1 (1971): 35–57.

94 *Some three hundred years earlier, in* Leviathan, *Hobbes . . .* See Thomas Hobbes, *Leviathan*, Parts I and II, Revised Edition, eds. A. P. Martinich and Brian Battiste (Peterborough, ON: Broadview Press, 2010).

4. The Wisdom of Psychopaths

101 *Back in 2010, Jonason . . .* See Peter K. Jonason, Norman P. Li, and Emily A. Teicher, "Who Is James Bond? The Dark Triad as an Agentic Social Style," *Individual Differences Research* 8, no. 2 (2010): 111–20.

102 *Jonason's study saw two hundred college students* . . . See P. K. Jonason, N. P. Li, Gregory W. Webster, and David P. Schmitt, "The Dark Triad: Facilitating a Short-Term Mating Strategy in Men," *European Journal of Personality* 23 (2009): 5–18, doi:10.1002/per.698.

103 *A 2005 study, conducted by a joint team of psychologists and neuroeconomists* . . . See Baba Shiv, George Loewenstein, and Antoine Bechara, "The Dark Side of Emotion in Decision-Making: When Individuals with Decreased Emotional Reactions Make More Advantageous Decisions," *Cognitive Brain Research* 23, no. 1 (2005): 85–92, doi:10.1016/j.cogbrainres.2005.01.006. Neuroeconomics is an interdisciplinary field focusing on the mental processes that underlie financial decision making. It combines research methods from neuroscience, economics, and social and cognitive psychology, as well as incorporating ideas and concepts from theoretical biology, computer science, and mathematics. For those who wish to explore the relationship between emotions and decision making in greater detail, an excellent place to start is Antonio Damasio's laudably readable *Descartes' Error: Emotion, Reason, and the Human Brain* (New York: Putnam, 1994). It should also be noted that the results of Shiv et al.'s experiment do not negate the fact that emotions often *do* play a useful role in financial decision making, and that unchecked risk taking (as if, in the current financial climate, we need any reminding) can sometimes lead to disaster. To illustrate, while the brain-damaged players did well in the specific game in the study, *outside* the lab they didn't fare so well—three of them, for example, having filed for personal bankruptcy. Their inability to experience fear led to excess risk taking in the real world, and their lack of emotional judgment sometimes led them into the clutches of people who took advantage of them. The bottom line is that while emotions can undoubtedly sometimes get in the way of rational decision making, they do, nevertheless, play an important role in safeguarding our interests.

104 *A study conducted by economist Cary Frydman and his colleagues* . . . See Cary Frydman, Colin Camerer, Peter Bossaerts, and Antonio Rangel, "MAOA-L Carriers Are Better at Making Optimal Financial Decisions Under Risk," *Proceedings of the Royal Society B* 278, no. 1714 (2011): 2053–59, doi:10.1098/rspb.2010.2304. On the link between the "warrior gene" and aggression, Antonio Rangel, who heads the lab where Cary Frydman is based, urges caution. "Previous studies that have associated *MAOA-L* with aggression or impulsivity might have to be interpreted carefully," he points out. "The key question is whether, in the context of the lives of the subjects, these decisions were optimal or not." (See Debora McKenzie, "People with 'Warrior Gene' Better at Risky Decisions," *New Scientist*, December 9, 2010. www.newscientist.com/article/dn19830-people-with-warrior-gene-better-at-risky-decisions.html.) In a study published in 2009, for instance, Dominic Johnson, of the University of Edinburgh, found that *MAOA-L* carriers were indeed more aggressive, but only after a great deal of provocation, and without apparent impulsiveness—a finding, like Frydman's, which seems to hint more at strategic self-interest

than indiscriminate self-destruction. See Rose McDermott, Dustin Tingley, Jonathan Cowden, Giovanni Frazzetto, and Dominic D. P. Johnson, "Monoamine Oxidase A Gene (MAOA) Predicts Behavioral Aggression Following Provocation," *PNAS* 106, no. 7 (2009): 2118–23, doi:10.1073/pnas.0808376106.

105 *"Contrary to previous discussion in the literature . . ."* See Richard Alleyne, "Gene That Makes You Good at Taking Risky Decisions," *The Telegraph*, December 8, 2010, www.telegraph.co.uk/science/science-news/8186570/Gene-that-makes-you-good-at-taking-risky-decisions.html.

105 *Additional support comes from work carried out by Bob Hare and his colleagues . . .* See: Paul Babiak, Craig S. Neumann, and Robert D. Hare, "Corporate Psychopathy: Talking the Walk," *Behavioral Sciences and the Law* 28, no. 2 (2010): 174–93, doi:10.1002/bsl.925.

106 *"The psychopath has no difficulty dealing with the consequences of rapid change . . ."* Quote taken from Alan Deutschman, "Is Your Boss a Psychopath?" *Fast Company*, July 1, 2005, www.fastcompany.com/magazine/96/open_boss.html.

107 *I caught up with him in the bar of a five-star hotel in New Orleans . . .* See Kevin Dutton, *Split-Second Persuasion: The Ancient Art and New Science of Changing Minds* (New York: Houghton Mifflin Harcourt, 2011).

109 *Research shows that one of the best ways of getting people to tell you about themselves . . .* See Morgan Worthy, Albert L. Gary, and Gay M. Kahn, "Self-Disclosure as an Exchange Process," *Journal of Personality and Social Psychology* 13, no. 1 (1969): 59–63.

109 *Research also shows that if you want to stop someone from remembering something . . .* See John Brown, "Some Tests of the Decay Theory of Immediate Memory," *Quarterly Journal of Experimental Psychology* 10, no. 1 (1958): 12–21, doi:10.1080/17470215808416249; Lloyd R. Peterson and Margaret J. Peterson, "Short-Term Retention of Individual Verbal Items," *Journal of Experimental Psychology* 58, no. 3 (1959): 193–98.

109 *And in clinical psychology, there comes a point in virtually every therapeutic intervention . . .* For more on the various techniques of therapeutic intervention, and on the work of Stephen Joseph, see Stephen Joseph, *Theories of Counselling and Psychotherapy: An Introduction to the Different Approaches*, 2nd revised edition (New York: Palgrave Macmillan, 2010).

110 *If, under certain conditions, psychopathy really is beneficial . . .* See Eyal Aharoni and Kent A. Kiehl, "Quantifying Criminal Success in Psychopathic Offenders," conference proceedings of the Society for the Scientific Study of Psychopathy, Montreal, Canada, May 2011.

111 *Helinä Häkkänen-Nyholm, a psychologist at the University of Helsinki . . .* See Helinä Häkkänen-Nyholm and Robert D. Hare, "Psychopathy, Homicide, and the Courts: Working the System," *Criminal Justice and Behavior* 36, no. 8 (2009): 761–77, doi:10.1177/0093854809336946.

111 *Remorse aside, Porter wondered . . .* See Stephen Porter, Leanne ten Brinke,

Alysha Baker, and Brendan Wallace, "Would I Lie to You? 'Leakage' in Deceptive Facial Expressions Relates to Psychopathy and Emotional Intelligence," *Personality and Individual Differences*, 51, no. 2 (2011): 133–37, doi:10.1016/j.paid.2011.03.031.

112 *Karim and his team at the University of Tübingen* . . . See Ahmed A. Karim, Markus Schneider, Martin Lotze, Ralf Veit, Paul Sauseng, Christoph Braun, and Niels Birbaumer. "The Truth About Lying: Inhibition of the Anterior Prefrontal Cortex Improves Deceptive Behavior," *Cerebral Cortex* 20, no. 1 (2010): 205–13, doi:10.1093/cercor/bhp090.

113 *and recent analysis using diffusion tensor imaging (DTI)* . . . See Michael C. Craig, Marco Catani, Quinton Deeley, Richard Latham, Eileen Daly, Richard Kanaan, Marco Picchioni, Philip K. McGuire, Thomas Fahy, and Declan G. M. Murphy, "Altered Connections on the Road to Psychopathy," *Molecular Psychiatry* 14 (2009): 946–53.

115 *Raine compared the performance of psychopaths and non-psychopaths on a simple learning task* . . . See Angela Scerbo, Adrian Raine, Mary O'Brien, Cheryl-Jean Chan, Cathy Rhee, and Norine Smiley, "Reward Dominance and Passive Avoidance Learning in Adolescent Psychopaths," *Journal of Abnormal Child Psychology* 18, no. 4 (1990): 451–63, doi:10.1007/BF00917646.

115 *Researchers at Vanderbilt University have delved a little deeper* . . . See Joshua W. Buckholtz, Michael T. Treadway, Ronald L. Cowan, Neil D. Woodward, Stephen D. Benning, Rui Li, M. Sib Ansari, et al., "Mesolimbic Dopamine Reward System Hypersensitivity in Individuals with Psychopathic Traits," *Nature Neuroscience* 13, no. 4 (2010): 419–21, doi:10.1038/nn.2510.

116 *"There has been a long tradition of research on psychopathy . . ."* For the full quote, and more details of the study, see "Psychopaths' Brains Wired to Seek Rewards, No Matter the Consequences," *Science Daily*, March 14, 2010, www.sciencedaily.com/releases/2010/03/100314150924.htm.

116 *Jeff Hancock, professor of computing and information science at Cornell* . . . See Jeffrey T. Hancock, Michael T. Woodworth, and Stephen Porter, "Hungry Like the Wolf: A Word-Pattern Analysis of the Language of Psychopaths," *Legal and Criminological Psychology* (2011), doi:10.1111/j.2044-8333.2011.02025.x.

118 *Fecteau and her coworkers used TMS to stimulate the somatosensory cortex* . . . See Shirley Fecteau, Alvaro Pascual-Leone, and Hugo Théoret, "Psychopathy and the Mirror Neuron System: Preliminary Findings from a Non-Incarcerated Sample," *Psychiatry Research* 160, no. 2 (2008): 137–44.

118 . . . *the work of highly specialized, and aptly named, brain structures called mirror neurons* . . . Mirror neurons were first discovered (in monkeys) in 1992 by an Italian research team led by Giacomo Rizzolatti, at the University of Parma. Put simply, they are brain cells specifically equipped to mimic the actions—and feelings—of others. See: Giuseppe Di Pellegrino, Luciano Fadiga, Leonardo Fogassi, Vittorio Gallese, and Giacomo Rizzolatti," Understanding Motor Events: A Neurophysiological Study," *Experimental Brain Research* 91 (1992):

176–80; G. Rizzolatti, L. Fadiga, V. Gallese, and L. Fogassi, "Premotor Cortex and the Recognition of Motor Actions," *Cognitive Brain Research* 3 (1996): 131–41.

118 *yawn contagion* . . . For an interesting recent paper on yawn contagion and empathy, see Ivan Norscia and Elisabetta Palagi, "Yawn Contagion and Empathy in *Homo sapiens*," *PLoS ONE* 6, no. 12 (2011): e28472, doi:10.1371/journal.pone.0028472.

119 *In an emotion recognition task using fMRI* . . . See Heather L. Gordon, Abigail A. Baird, and Alison End, "Functional Differences Among Those High and Low on a Trait Measure of Psychopathy," *Biological Psychiatry* 56, no. 7 (2004): 516–21.

120 *Yawei Cheng, at the National Yang-Ming University* . . . See Yawei Cheng, Ching-Po Lin, Ho-Ling Liu, Yuan-Yu Hsu, Kun-Eng Lim, Daisy Hung, and Jean Decety, "Expertise Modulates the Perception of Pain in Others," *Current Biology* 17, no. 19 (2007): 1708–13, doi:10.1016/j.cub.2007.09.020.

121 *the Trier* . . . See Clemens Kirschbaum, Karl-Martin Pirke, and Dirk H. Hellhammer, "The Trier Social Stress Test—A Tool for Investigating Psychobiological Stress Responses in a Laboratory Setting," *Neuropsychobiology* 28, no. 1–2 (1993): 76–81.

123 *Ray postulated an inverted-U-shaped function* . . . See John J. Ray and J.A.B. Ray, "Some Apparent Advantages of Subclinical Psychopathy," *Journal of Social Psychology* 117 (1982): 135–42.

123 *"Both extremely high and extremely low levels of psychopathy . . ."* Ray and Ray, 1982.

124 *Hare and Babiak have developed an instrument called the Business Scan* . . . For more on the B-Scan, see www.b-scan.com/index.html (accessed February 3, 2012). For an entertaining and accessible introduction to psychopathy in corporate settings, see Paul Babiak and Robert D. Hare, *Snakes in Suits: When Psychopaths Go to Work* (New York: HarperBusiness, 2006).

5. Make Me a Psychopath

130 *"And I read a report the other day that linked a significant rise in the number of all-female gangs* . . ." To get a flavor of what Hare is talking about, see Tom Geoghegan, *BBC News Magazine*. May 5, 2008, http://news.bbc.co.uk/1/hi/magazine/7380400.stm. For a more academic slant on things, see Susan Batchelor, "Girls, Gangs, and Violence: Assessing the Evidence," *Probation Journal* 56, no. 4 (2009): 399–414, doi: 10.1177/0264550509346501.

130 *Harvard psychologist Steven Pinker has recently flagged this* . . . See Steven Pinker, *The Better Angels of Our Nature: Why Violence Has Declined* (New York: Viking, 2011).

130 *Trawling through the court records of a number of European countries* . . See Manuel Eisner, "Long-Term Historical Trends in Violent Crime," *Crime and Justice* 30 (2003): 83–142.

131 *Similar patterns have elsewhere been documented* . . . Michael Shermer, "The Decline of Violence," *Scientific American*, October 7, 2011, www.scientificam erican.com/article.cfm?id=the-decline-of-violence.

131 *The same goes for war* . . . Pinker, *The Better Angels of Our Nature*, 47–56: "Rates of Violence in State and Nonstate Societies."

131 *"Beginning in the eleventh or twelfth [century], and maturing in the seventeenth and eighteenth* . . ." Quote taken from Shermer, "The Decline of Violence," 2011. Be skeptical of claims that we live in an ever more dangerous world.

133 *"With infinitely more complex securities* . . ." Quote taken from Gary Strauss, "How Did Business Get So Darn Dirty?" *USA Today* (Money), June 12, 2002, www.usatoday.com/money/covers/2002-06-12-dirty-business.htm.

133 *and in a recent issue of the* Journal of Business Ethics . . . See Clive R. Boddy, "The Corporate Psychopaths Theory of the Global Financial Crisis," *Journal of Business Ethics* 102, no. 2 (2011): 255–59, doi:10.1007/s10551-011-0810-4. (The moniker "corporate Attila" was first applied to Fred "the Shred" Goodwin, who, as CEO of the Royal Bank of Scotland from 2001–2009, racked up a corporate loss of £24.1 billion, the highest in U.K. history.)

134 *On the other hand, however, there's society in general, proclaims Charles Elson* . . . See Strauss, "How Did Business Get So Darn Dirty?"

135 *"Ms. Smart overcame it. Survived it. Triumphed over it* . . ." For coverage of this quote in the media, see Camille Mann, "Elizabeth Smart Was Not Severely Damaged by Kidnapping, Defense Lawyers Claim," *CBS News*, May 19, 2011, www.cbsnews.com/8301-504083_162-20064372-504083.html.

136 *In a recent study by the Crime and Justice Centre at King's College, London* . . . For an in-depth analysis of youth crime in the U.K., including prevalence, motivation, and risk factors, see Debbie Wilson, Clare Sharp, and Alison Patterson, "Young People and Crime: Findings from the 2005 Offending, Crime and Justice Survey" (London: Home Office, 2005).

136 *If the results of a recent study by Sara Konrath* . . . Sara Konrath, Edward H. O'Brien, and Courtney Hsing, "Changes in Dispositional Empathy in American College Students over Time: A Meta-Analysis," *Personality and Social Psychology Review* 15, no. 2 (2011): 180–98, doi:10.1177/1088868310377395.

136 *the Interpersonal Reactivity Index* . . . For the background to, and development of, the IRI, see Mark H. Davis, "A Multidimensional Approach to Individual Differences in Empathy," *JSAS Catalog of Selected Documents in Psychology* 10, no. 85 (1980); and M. H. Davis, "Measuring Individual Differences in Empathy: Evidence for a Multidimensional Approach," *Journal of Personality and Social Psychology* 44, no. 1 (1983): 113–26.

136 *"College kids today are about 40 percent lower in empathy* . . ." See "Today's College Students More Likely to Lack Empathy," *U.S. News* (Health), May 28, 2010, http://health.usnews.com/health-news/family-health/brain-and-behav ior/articles/2010/05/28/todays-college-students-more-likely-to-lack-empathy.

136 *More worrying still, according to Jean Twenge . . .* See Jean M. Twenge, Sara Konrath, Joshua D. Foster, W. Keith Campbell, and Brad J. Bushman, "Egos Inflating Over Time: A Cross-Temporal Meta-Analysis of the Narcissistic Personality Inventory," *Journal of Personality* 76, no. 4 (2008a): 875–901, doi:10.1111/j.1467-6494.2008.00507.x; Twenge et al., "Further Evidence of an Increase in Narcissism Among College Students," *Journal of Personality* 76 (2008b): 919–27, doi:10.1111/j.1467-6494.2008.00509.x.

137 *Many people see the current group of college students . . .* See *U.S. News*, "Today's College Students More Likely to Lack Empathy."

137 *"People haven't had the same exposure to traditional values . . ."* See Thomas Harding, "Army Should Provide Moral Education for Troops to Stop Outrages," *The Telegraph*, February 22, 2011, www.telegraph.co.uk/news/8341030/Army-should-provide-moral-education-for-troops-to-stop-outrages.html.

137 *But the beginnings of an even more fundamental answer may lie . . .* See Nicole K. Speer, Jeremy R. Reynolds, Khena M. Swallow, and Jeffrey M. Zacks, "Reading Stories Activates Neural Representations of Perceptual and Motor Experiences," *Psychological Science* 20, no, 8 (2009): 989–99.

138 *Makes us, as Nicholas Carr puts it in his recent essay . . .* Nicholas Carr's "The Dreams of Readers" appears in Mark Haddon (ed.), *Stop What You're Doing and Read This!* (London: Vintage, 2011), a collection of essays about the transformative power of reading.

138 *The quicksilver virtual world . . .* Christina Clark, Jane Woodley, and Fiona Lewis, *The Gift of Reading in 2011: Children and Young People's Access to Books and Attitudes Towards Reading*—see www.literacytrust.org.uk/assets/0001/1303/The_Gift_of_Reading_in_2011.pdf.

138 *. . . we've been talking about the emergence of neurolaw . . .* For an excellent introduction to the emerging subdiscipline of neurolaw, see David Eagleman, "The Brain on Trial," *The Atlantic.*, July/August 2011, www.theatlantic.com/magazine/print/2011/07/the-brain-on-trial/8520/.

138 *The watershed study was published in 2002 . . .* See Avshalom Caspi, Joseph McClay, Terrie E. Moffitt, Jonathan Mill, Judy Martin, Ian W. Craig, Alan Taylor, and Richie Poulton, "Role of Genotype in the Cycle of Violence in Maltreated Children," *Science* 297, no. 5582 (2002): 851–54, doi:10.1126/science.1072290.

139 *The implications of the discovery have percolated into the courtroom . . .* For a nuanced discussion of the research, and controversy, surrounding the "warrior gene," see Ed Yong, "Dangerous DNA: The Truth About the 'Warrior Gene,'" *New Scientist*, April 12, 2010, www.newscientist.com/article/mg20627557.300-dangerous-dna-the-truth-about-the-warrior-gene.html?page=1.

139 *In 2006, Bradley Waldroup's defense attorney . . .* For more on the Waldroup case, see "What Makes Us Good or Evil?" BBC Horizon, September 7, 2011, www.youtube.com/watch?v=xmAyxpAFS1s. For more on the neural, genetic,

and psychological profiles of violent killers, listen to Barbara Bradley Hagerty's excellent series *Inside the Criminal Brain*, NPR, June 29–July 1, 2010, www.npr.org/series/128248068/inside-the-criminal-brain.

140 *The subject of neurolaw came up in the context of a wider discussion . . .* For more on the emerging field of cultural neuroscience, see Joan Y. Chiao and Nalini Ambady, "Cultural Neuroscience: Parsing Universality and Diversity across Levels of Analysis," in Shinobu Kitayama and Dov Cohen, eds., *Handbook of Cultural Psychology* (New York: Guilford Press, 2007), 237–54; and Joan Y. Chiao, ed., *Cultural Neuroscience: Cultural Influences on Brain Function*, Progress in Brain Research (New York: Elsevier, 2009).

140 *a hot new offshoot from the field of mainstream genetics . . .* For a clear and accessible introduction to the field of epigenetics, see Nessa Carey, *The Epigenetics Revolution: How Modern Biology Is Rewriting Our Understanding of Genetics, Disease, and Inheritance* (New York: Columbia University Press, 2012).

141 *Hare tells me about a study conducted in Sweden back in the 1980s . . .* See Gunnar Kaat, Lars O. Bygren, and Sören Edvinsson, "Cardiovascular and Diabetes Mortality Determined by Nutrition During Parents' and Grandparents' Slow Growth Period," *European Journal of Human Genetics* 10, no. 11 (2002): 682–88, doi:10.1038/sj.ejhg.5200859.

142 *"There was a writer back in the sixties," Alan Harrington . . .* See: Alan Harrington, *Psychopaths* (New York: Simon & Schuster, 1972).

142 *"Did I tell you about [this] paper which shows that people with high testosterone levels . . ."* See Robert A. Josephs, Michael J. Telch, J. Gregory Hixon, Jacqueline J. Evans, Hanjoo Lee, Valerie S. Knopik, John E. McGeary, Ahmad R. Hariri, and Christopher G. Beevers, "Genetic and Hormonal Sensitivity to Threat: Testing a Serotonin Transporter Genotype X Testosterone Interaction," doi:10.1016/j.psyneuen.2011.09.006.

142 *As an amusing backdrop, "Gary Gilmore's Eyes" by the Adverts is playing . . .* See "Gary Gilmore's Eyes" / "Bored Teenagers" (August 19, 1977: Anchor Records ANC1043).

144 *Transcranial magnetic stimulation (or TMS) was first developed by Dr. Anthony Barker . . .* For the inaugural study using TMS, see Anthony T. Barker, Reza Jalinous, and Ian L. Freeston, "Non-Invasive Magnetic Stimulation of Human Motor Cortex," *Lancet* 325, no. 8437 (1985): 1106–07, doi:10.1016/S0140-6736(85)92413-4.

145 *Indeed, Liane Young and her team at MIT . . .* See Liane Young, Joan Albert Camprodon, Marc Hauser, Alvaro Pascual-Leone, and Rebecca Saxe, "Disruption of the Right Temporoparietal Junction with Transcranial Magnetic Stimulation Reduces the Role of Beliefs in Moral Judgments," *PNAS* 107, no. 15 (2010): 6753–58, doi:10.1073/pnas.0914826107. Imagine you observe an employee at a chemical plant pouring some sugar into a work colleague's cup of coffee. The sugar is stored in a container marked "toxic." As you watch, a crack

in time suddenly opens up, and out of it, in a dodgy puff of smoke, an ethereal moral philosopher appears, in a hazard suit and goggles, and presents you with four scenarios. These scenarios incorporate two independent dimensions of possibility space, aligned orthogonally to one another. The first dimension relates to what the employee *believes* the contents of the container to be (sugar or toxic powder.) The second dimension maps onto what, in actuality, the container really *does* contain (sugar or toxic powder). So we have, in effect, the following mélange of quantum possibility, distilled from a cocktail of outcome and personal belief (see the figure below):

1. The employee thinks that the powder is sugar. And it is indeed sugar. The colleague drinks the coffee. And survives.
2. The employee thinks that the powder is sugar. But it is, in fact, toxic. The colleague drinks the coffee. And dies.
3. The employee thinks that the powder is toxic. But it's sugar. The colleague drinks the coffee. And survives.
4. The employee thinks that the powder is toxic. And, you guessed it, it is indeed toxic. The colleague drinks the coffee. And dies.

BELIEF \ OUTCOME	Neutral	Negative
Neutral	1	2
Negative	3	4

Bearing in mind that, according to a basic tenet of criminal law, "the act does not make the person guilty unless the mind is also guilty," how permissible, asks the philosopher, on a scale of 1 to 7 (1 = completely forbidden; 7 = completely okay), would you rate the actions of the employee to be in each of these four scenarios?

In 2010, Liane Young, at the Department of Brain and Cognitive Sciences, Massachusetts Institute of Technology, and her coworkers asked volunteers to make precisely these judgments, as part of an investigation into the neurobiology of moral decision making.

But there was a catch.

Prior to making their judgments, some of the participants in the study received transcranial magnetic stimulation (TMS) to a region of the brain known to be associated with moral processing (the right temporoparietal junction, or RTPJ). More specifically (which makes it different from the morality zapping that Ahmed Karim was up to in chapter 4), moral processing when evaluating the beliefs, attitudes, and intentions underlying the actions of a third party.

Would this artificial stimulation of participants' RTPJ affect the way they viewed the different scenarios? Young and her coauthors wondered. Was morality, in other words, malleable?

The answer, it transpired, was yes.

When the moral judgments of the experimental group were compared to those of a corresponding group that received TMS at a control site (i.e., not at the RTPJ), Young detected a pattern. In Scenario 3 (where harm is intended, but the outcome turns out to be positive), those participants who received TMS in their RTPJ judged the action of the employee as more morally permissible than those who'd received it elsewhere.

Morality, it appears, really can be manipulated. Or rather, a component of it can be. The ability to accurately ascribe intentionality in judging the behavior of another may be ratcheted up or down.

146 *In 1993, in a book that bore their name . . .* See Andy McNab, *Bravo Two Zero: The Harrowing True Story of a Special Forces Patrol Behind the Lines in Iraq* (London: Bantam Press, 1993; New York: Dell, 1994).

147 *The show is called* Extreme Persuasion . . . To listen to the show, go to www.bbc.co.uk/programmes/p006dg3y.

148 *And like most things in the regiment, there's a good reason for that . . .* For more on Andy McNab's experiences in the Special Air Service, including its fearsome selection process and interrogation techniques, see Andy McNab, *Immediate Action: The Explosive True Story of the Toughest—and Most Highly Secretive—Strike Force in the World* (London: Bantam Press, 1995; New York: Dell, 1996).

6. The Seven Deadly Wins

162 *after launching the Great British Psychopath Survey . . .* To listen to the show, go to http://soundcloud.com/profkevindutton/great-british-psychopath.

162 *Participants were directed onto my website . . .* See Michael R. Levenson, Kent A. Kiehl, and Cory M. Fitzpatrick, "Assessing Psychopathic Attributes in a Noninstitutionalized Population," *Journal of Personality and Social Psychology* 68, no. 1 (1995): 151–58.To take the test, go to www.kevindutton.co.uk/.

163 *The Paddock Centre . . .* Psychopaths are notoriously difficult to treat, and their charm and persuasive skills often quite literally flatter to deceive—giving the impression that progress has been made, when, in fact, the psychopath is faking rehabilitation in order (in most cases) to gain parole. Recently, however, a

new treatment for intractable juvenile offenders with psychopathic tendencies has provided cause for optimism. Michael Caldwell, a psychologist at the Mendota Juvenile Treatment Center in Madison, Wisconsin, has had promising results with an intensive one-on-one therapeutic technique known as decompression: the aim of which is to end the vicious cycle in which punishment for bad behavior inspires more bad behavior, which is then, in turn, punished again . . . and so on, and so on . . . Over time, the behavior of the incarcerated youths treated by Caldwell became gradually more manageable, with the result that they were subsequently able to participate in more mainstream rehabilitation services. To illustrate, a group of more than 150 youths enrolled in Caldwell's program were 50 percent less likely to engage in violent crime following treatment than a comparable group who underwent rehabilitation at regular juvenile correctional facilities.

For more information on decompression and the treatment of psychopaths in general, see Michael F. Caldwell, Michael Vitacco, and Gregory J. Van Rybroek, "Are Violent Delinquents Worth Treating? A Cost-Benefit Analysis," *Journal of Research in Crime and Delinquency* 43, no. 2 (2006): 148–68, doi: 10.1177/0022427805280053.

174 *Take Steve Jobs*. . . See: John Arlidge, "A World in Thrall to the iTyrant," Sunday *Times* News Review, October 9, 2011.

175 *James Rilling, associate professor of anthropology at Emory University* . . . See James K. Rilling, Andrea L. Glenn, Meeta R. Jairam, Giuseppe Pagnoni, David R. Goldsmith, Hanie A. Elfenbein, and Scott O. Lilienfeld, "Neural Correlates of Social Cooperation and Non-Cooperation as a Function of Psychopathy," *Biological Psychiatry* 61, no. 11 (2007): 1260–71.

176 *Lee Crust and Richard Keegan, at the University of Lincoln* . . . See Crust and Keegan, "Mental Toughness and Attitudes to Risk-Taking," *Personality and Individual Differences* 49, no. 3 (2010): 164–68.

180 *Mark Williams, professor of clinical psychology* . . . Williams and his team are based in the Oxford Mindfulness Centre, University of Oxford. Learn more about current research at the center by visiting its website: http://oxfordmindfulness.org/. For those interested in reading about mindfulness, see also Mark Williams and Danny Penman, *Mindfulness: A Practical Guide to Finding Peace in a Frantic World* (London: Piatkus, 2011; New York: Rodale Books, 2011).

181 *What, precisely, is it that makes a successful trader?* Reams have been written in answer to this question. For those who wish to keep things light, and inject a pinch of fantasy into the mix, I strongly recommend a novel by Robert Harris: *The Fear Index* (New York: Knopf, 2012).

184 . . . *the "imagined" potential reality is significantly more discomfiting* . . . See Artur Z. Arthur, "Stress as a State of Anticipatory Vigilance," *Perceptual and Motor Skills* 64, no. 1 (1987): 75–85, doi:10.2466/pms.1987.64.1.75.

7. Supersanity

188 *Harrington cites some examples . . .* See Alan Harrington, *Psychopaths* (New York: Simon & Schuster, 1972), 45.

189 *"What he [the psychopath] believes he needs to protest against . . ."* See Cleckley, *The Mask of Sanity* (St. Louis, MO: C. V. Mosby, 1941, 1976), 391, www.cassiopaea.org/cass/sanity_1.pdf.

189 *"[The psychopath] is an elite with the potential ruthlessness of an elite . . ."* See Norman Mailer, *The White Negro,* first published in *Dissent* (Fall 1957), www.learntoquestion.com/resources/database/archives/003327.html.

189 *"whether we want to admit it or not . . ."* See Harrington, *Psychopaths,* 233.

190 *Saint Paul, as we know him now . . .* For a detailed biography of Saint Paul and informed insights into his complex psychology, see A. N. Wilson, *Paul: The Mind of the Apostle* (New York: W. W. Norton, 1997).

191 *"a total failure of political bravado . . ."* See L. Michael White, *From Jesus to Christianity: How Four Generations of Visionaries and Storytellers Created the New Testament and Christian Faith* (San Francisco: HarperCollins, 2004), 170.

192 *"If you can meet with Triumph and Disaster . . ."* "If—," the poem by Rudyard Kipling from which these lines are taken, first appeared in his collection *Rewards and Fairies* (London: Macmillan, 1910).

192 *Derek Mitchell at University College, London . . .* See Derek G. V. Mitchell, Rebecca A. Richell, Alan Leonard, and James R. Blair, R. "Emotion at the Expense of Cognition: Psychopathic Individuals Outperform Controls on an Operant Response Task," *Journal of Abnormal Psychology* 115, no. 3 (2006): 559–66.

195 *In fact, this ability to concentrate purely on the task in hand . . .* For more on the concept of flow, see Mihály Csíkszentmihályi, *Finding Flow: The Psychology of Engagement with Everyday Life* (New York: Basic Books, 1996).

195 *In 2011, Martin Klasen at Aachen University discovered that moments of flow . . .* Martin Klasen, Rene Weber, Tilo T. J. Kircher, Krystyna A. Mathiak, and Klaus Mathiak, "Neural Contributions to Flow Experience During Video Game Playing," *Social Cognitive and Affective Neuroscience* 7, no. 4 (2012): 485–95, doi: 10.1093/scan/nsr021.

195 *In the same year that Klasen was playing video games . . .* See Elsa Ermer, Joshua D. Greene, Prashanth K. Nyalakanti, and Kent A. Kiehl, "Abnormal Moral Judgments in Psychopathy," poster presented at the Society for the Scientific Study of Psychopathy Conference, Montreal, Canada, May 2011.

198 *The neuroscientist Richard Davidson, at the University of Wisconsin . . .* See Antoine Lutz, Lawrence L. Greischar, Nancy B. Rawlings, Matthieu Ricard, and Richard J. Davidson, "Long-Term Meditators Self-Induce High-Amplitude Gamma Synchrony During Mental Practice," *PNAS* 101, no. 46 (2004): 16369–73, doi:10.1073/pnas.0407401101.

Richard Davidson is the director of the Laboratory for Affective Neuroscience at the University of Wisconsin. To find out more about his work, visit the lab's website at: http://psyphz.psych.wisc.edu. See also Richard J. Davidson and Sharon Begley, *The Emotional Life of Your Brain: How Its Unique Patterns Affect the Way You Think, Feel, and Live—And How You Can Change Them* (New York: Hudson Street Press, 2012).

198 *"There is a lot of evidence [to suggest] that the best sportsmen and women . . ."* Quote taken from Steve Conner, "Psychology of Sport: How a Red Dot Swung It for Open Champion," *The Independent*, July 20, 2010, www.independent.co.uk/sport/general/others/psychology-of-sport-how-a-red-dot-swung-it-for-open-champion-2030349.html.

199 *"The mind is deliberately kept at the level of* bare attention . . ." See Bikkhu Bodhi, "Right Mindfulness (Samma Sati)," chapter 6 in *The Noble Eightfold Path: The Way to the End of Suffering* (Onalaska, WA: BPS Pariyatti Publishing, 2000).

199 *According to the Mahāsatipaṭṭhāna Sutta . . .* See *Mahāsatipaṭṭhāna Sutta: The Great Discourse on the Establishing of Awareness* (Onalaska, WA: Vipassana Research Publications, 1996).

200 *"The first component [of mindfulness] involves the self-regulation of attention . . ."* See Scott R. Bishop, Mark Lau, Shauna Shapiro, Linda Carlson, Nicole D. Anderson, James Carmody, Zindel V. Segal, et al., "Mindfulness: A Proposed Operational Definition," *Clinical Psychology: Science and Practice* 11, no. 3 (2004): 230–41, doi:10.1093/clipsy.bph077.

200 *"In the beginner's mind there are many possibilities . . ."* See Shunryu Suzuki, *Zen Mind, Beginner's Mind: Integrated Talks on Zen Meditation and Practice*, Trudy Dixon and Richard Baker, eds. (New York and Tokyo: Weatherhill, 1970).

202 *Psychopaths, it appears, far from being callous and unemotional . . .* Mehmet Mahmut and Louise Cridland, "Exploring the Relationship Between Psychopathy and Helping Behaviors," poster presented at the Society for the Scientific Study of Psychopathy Conference, Montreal, Canada, May 2011.

204 *"It is pleasure that lurks in the practice of every one of your virtues . . ."* See W. Somerset Maugham, *Of Human Bondage* (London: George H. Doran and Company, 1915).

204 *As a case in point, Diana Falkenbach and Maria Tsoukalas . . .* Falkenbach and Tsoukalas, "Can Adaptive Psychopathic Traits Be Observed in Hero Populations?" Poster presented at the Society for the Scientific Study of Psychopathy Conference. Montreal, Canada, May 2011.

205 *Philip Zimbardo, founder of the Heroic Imagination Project . . .* To find out more about the Heroic Imagination Project, visit its website at http://heroicimagination.org/.

205 *In 1971, in an experiment which has long since been inaugurated . . .* Philip G. Zimbardo, "The Power and Pathology of Imprisonment," Hearings before Subcommittee No. 3 of the Committee on the Judiciary, House of Representatives, Ninety-Second Congress, First Session on Corrections, Part II, Prisons,

Prison Reform and Prisoner's Rights: California, *Congressional Record*, Serial No. 15, October 25, 1971. Washington, DC: U.S. Government Printing Office, 1971, www.prisonexp.org/pdf/congress.pdf.

207 *Rationalizations included . . .* See Irving L. Janis, *Groupthink: A Psychological Study of Policy Decisions and Fiascoes*, 2nd ed. (Boston: Houghton Mifflin, 1982).

207 *Such a position would better align itself . . .* See Timothy A. Judge, Beth A. Livingston, and Charlice Hurst, "Do Nice Guys—and Gals—Really Finish Last? The Joint Effects of Sex and Agreeableness on Income," *Journal of Personality and Social Psychology* 102, no. 2 (2012): 390–407, doi:10.1037/a0026021.

208 *Studies have revealed that when psychopaths are shown distressing images . . .* See Uma Vaidyanathan, Jason R. Hall, Christopher J. Patrick, and Edward M. Bernat, "Clarifying the Role of Defensive Reactivity Deficits in Psychopathy and Antisocial Personality Using Startle Reflex Methodology," *Journal of Abnormal Psychology* 120, no. 1 (2011): 253–58, doi:10.1037/a0021224.

209 *But neither of those two scenarios fitted this guy's profile . . .* For those interested in finding out more about the science of criminal profiling, see Brent Turvey, *Criminal Profiling: An Introduction to Behavioral Evidence Analysis* (San Diego: Academic Press, 1999); David V. Canter and Laurence J. Alison, eds., *Criminal Detection and the Psychology of Crime* (Brookfield, VT: Ashgate Publishing, 1997).

211 *"But for another category of serial killer, those we call sadistic serial killers . . ."* See Andreas Mokros, Michael Osterheider, Stephen J. Hucker, and Joachim Nitschke, "Psychopathy and Sexual Sadism," *Law and Human Behavior* 35, no. 3 (2011): 188–99.

The Kelleher Typology for male serial killers differentiates such individuals into four discrete categories: Visionaries, Missionaries, Hedonists, and Power Seekers. Visionaries respond to psychic messages, divine communication, and/or influential alter egos commanding them to kill. Missionaries feel it incumbent upon themselves to "clean up" society, usually preying upon prostitutes or other selectively designated minority targets such as homosexuals, or particular ethnic or religious groups. Hedonists—the most common type of male serial killer—are predominantly pleasure-oriented, often getting a buzz from killing. They can be further subdivided into three distinct types: lust killers (who kill for sexual gratification), thrill killers (who kill purely for the pleasure of hunting and slaying their prey), and comfort killers (who kill for material gain). Finally, Power Seekers kill to exert control over their victims. Many such killers sexually abuse their victims, but differ from lust killers in that rape is used as a means of domination, as opposed to sexual gratification.

The Kelleher Typology for female serial killers comprises five different types: Black Widows, Angels of Death, Sexual Predators, Revenge Killers, and

Profit Killers. Black Widows kill family members, friends, and anyone with whom they have forged a close personal relationship, the primary aim being attention and sympathy. Angels of Death work in hospitals and nursing homes and are exhilarated by their own power over life and death, often bringing victims to the brink of death and then miraculously "curing" them. This type of killer is usually diagnosed with Munchausen Syndrome by Proxy.

The motives of Sexual Predators, Revenge Killers, and Profit Killers, respectively, are fairly self-evident—though it should be noted that Sexual Predators are extremely rare (with Aileen Wuornos arguably being pretty much the only example of a female serial killer fitting this description). In contrast, Profit Killers constitute the most common type of female serial killer, with almost three-quarters of such women falling into this category.

For those who wish to find out more about serial killers, of both the male and female variety, see Michael D. Kelleher and C.L. Kelleher, *Murder Most Rare: The Female Serial Killer* (Westport, CT: Praeger, 1998); and Michael Newton, *The Encyclopedia of Serial Killers*, 2nd ed. (New York: Checkmark Books, 2006).

211 *"The answer, I believe, is no . . ."* See Heinz Wimmer and Josef Perner, "Beliefs About Beliefs: Representation and Constraining Function of Wrong Beliefs in Young Children's Understanding of Deception," *Cognition* 13, no. 1 (1983): 103–28, doi:10.1016/0010-0277(83)90004-5.

212 *Beasley tells me about a study conducted by Alfred Heilbrun . . .* See Alfred B. Heilbrun, "Cognitive Models of Criminal Violence Based upon Intelligence and Psychopathy Levels," *Journal of Consulting and Clinical Psychology* 50, no. 4 (1982): 546–57.

213 *Paul Ekman, at the University of California, Berkeley . . .* The work of Paul Ekman and Robert Levenson is described in Daniel Goleman, "Dalai Lama" (foreword), *Destructive Emotions: How Can We Overcome Them? A Scientific Dialogue with the Dalai Lama* (New York: Bantam Books, 2003). For general background, see also Paul Ekman, Richard J. Davidson, Matthieu Ricard, and B. Alan Wallace, "Buddhist and Psychological Perspectives on Emotions and Well-Being," *Current Directions in Psychological Science* 14, no. 2 (2005): 59–63, doi:10.1111/j.0963-7214.2005.00335.x.

214 *Sabrina Demetrioff, at the University of British Columbia . . .* Data as yet unpublished.

216 *Chris Patrick of Florida State University compared the reactions of psychopaths and non-psychopaths . . .* See Christopher J. Patrick, Margaret M. Bradley, and Peter J. Lang, "Emotion in the Criminal Psychopath: Startle Reflex Modulation," *Journal of Abnormal Psychology* 102, no. 1 (1993): 82–92.

216 *The greatest worth, wrote the eleventh-century Buddhist teacher Atisha . . .* For an accessible guide to the writings, and philosophy, of Atisha, see Geshe Sonam Rinchen, *Atisha's Lamp for the Path to Enlightenment*, ed. and trans. by Ruth Sonam (Ithaca, NY: Snow Lion Publications, 1997).

216 *The latest FBI crime figures estimate . . .* See Blake Morrison, "Along Highways, Signs of Serial Killings," *USA Today*, October 5, 2010, www.usatoday.com/news/nation/2010-10-05-1Ahighwaykiller05_CV_N.htm.

220 *It's a poem about moths . . .* See Don Marquis, "the lesson of the moth," in *The Annotated Archy and Mehitabel*, ed. Michael Sims (New York: Penguin, 2006).

ACKNOWLEDGMENTS

Psychologically speaking, writers come in all different shapes and sizes. For me, writing a book that makes people laugh would, notionally, be relatively easy. Writing a book that makes people laugh and think—as I did previously with *Split-Second Persuasion* (or so I'm told by those who like me)—is a little more difficult. Writing a book that just makes people think, well . . . that doesn't come easy at all.

The Wisdom of Psychopaths arguably falls into this third category (though if, at times, I've managed to raise a smile, let's not fall out about it). Psychopaths, undeniably, are fascinating. But the plain unvarnished truth is that there is nothing remotely funny about them. They can be dangerous, destructive, and deadly—and any serious writer has a duty of care to handle them as judiciously on the printed page as they would were they to encounter them in real life.

Such scrupulous editorial hygiene is even more important under conditions of existential favor: when one advances the notion that the brain of the psychopath is not, in its entirety, the glacial, inhospitable world glimpsed, as is so often the case, in remote neurological orbit within the teeming synaptic firmament, but offers instead—contrary to popular belief—a habitable psychological refuge for normal, regular people during the course of their everyday lives (at least, that is, in its milder, more temperate regions). Evidence must be presented within hermetically sealed scientific argument, empirically sterilized to eradicate even the most infinitesimal microbes of hyperbole and glamorization, and conclusions generated under strictly controlled, highly secure conditions.

Yet psychopaths are as beguiling on paper as they are face-to-face, and my wife assures me that I have not escaped their devious psychological clutches unscathed. So naturally, she has a plan. To redress the balance and make it all up to her, my next book, she insists, should be a treatise on love and compassion—two attributes that in my considered opinion are wholly overrated and inestimably superfluous (fat chance of *that* book ever getting written, then). On which note, Elaine, I just want to say: Thanks for bloody nothing, love. You'll be hearing from my solicitor very soon.

Billy Wilder once said that agents are like car tires: to get anywhere at all, you need at least four of them, and they need to be rotated every five thousand miles. Personally, I cannot espouse enthusiastically enough the merits of the unicycle—in particular, those of the Patrick Walsh variety. With the aid of my specialized puncture-repair kit, Jake Smith-Bosanquet, I have ridden Patrick for a good few years now, and have relished every minute in the saddle. God knows where the next adventure will take us.

Other people without whose help this book would never have seen the dark of night (and who duly coughed up the inclusion fee) are as follows: Denis Alexander, Paul Babiak, Alysha Baker, Helen Beardsley, James Beasley III, Peter Bennett, James Blair, Michael Brooks, Alex Christofi, David Clark, Claire Conville, Nick Cooper, Sean Cunningham, Kathy Daneman, Ray Davies, Roger Deeble, Mariette DiChristina, Liam Dolan, Jennifer Dufton, Robin Dunbar, Elsa Ermer, Peter Fenwick, Simon Fordham, Mark Fowler, Susan Goldfarb, Graham Goodkind, Annie Gottlieb, Cathy Grossman, Robert Hare, Amelia Harvell, John Horgan, Glyn Humphreys, Hugh Jones, Terry Jones, Stephen Joseph, Larry Kane, Deborah Kent, Nick Kent, Paul Keyton, Kent Kiehl, Jennifer Lau, Scott Lilienfeld, Howard Marks, Tom Maschler, Matthias Matuschik, Andy McNab, Alexandra McNicoll, Drummond Moir, Helen Morrison, Joseph Newman, Richard Newman, Jonica Newby, Steven Pinker, Stephen Porter, Caroline Pretty, Philip Pullman, Martin Redfern, Christopher Richards, Ann Richie, Ruben Richie, Joe Roseman, John Rogers, Jose Romero-Urcelay, Tim Rostron, Debbie Schiesser, Henna Silvennoinen, Jeanette Slinger, Nigel

Stratton, Christine Temple, Leanne ten Brinke, John Timpane, Lisa Tuffin, Essi Viding, Dame Marjorie Wallace, Fraser Watts, Pete Wilkins, Mark Williams, Robin Williams, Andrea Woerle, Philip Zimbardo, and Konstantina Zougkou. (Note: Though of questionable importance anyway, Ian Collins did not see fit to stump up the required disbursement, and thus shall herewith remain conspicuous by his absence.)

Special thanks also go to my editors at William Heinemann, Tom Avery and Jason Arthur, and to the equally fastidious Amanda Moon and Karen Maine at Farrar, Straus and Giroux.

INDEX